Ernest Ingersoll

Old Ocean

Ernest Ingersoll

Old Ocean

ISBN/EAN: 9783337037918

Printed in Europe, USA, Canada, Australia, Japan

Cover: Foto ©berggeist007 / pixelio.de

More available books at **www.hansebooks.com**

OLD OCEAN

BY

ERNEST INGERSOLL

ILLUSTRATED

BOSTON

D. LOTHROP AND COMPANY

32 FRANKLIN STREET

CONTENTS.

I   146,000,000. SQUARE MILES OF WATER   .   7

II.   WAVES AND CURRENTS   25

III. EARLY VOYAGES, AND THE SOUTH POLE   .   45

IV.   THE POLAR REGIONS   63

V.   SHIPS AND THEIR RIGGING   79

VI.   WAR-SHIPS AND NAVAL BATTLES .   .   .   102

VII. ROBBERS OF THE SEA   119

VIII. THE MERCHANTS OF THE SEA   .   .   .   135

IX.   THE DANGERS OF THE DEEP   154

X.   LIFE UNDER THE WAVES   170

XI.   SEA ANIMALS — THE LESSER HALF   .   .   185

XII. SEA ANIMALS — THE GREATER HALF   .   .   204

LIST OF ILLUSTRATIONS.

	PAGE.
HAULING IN THE LIFE-SAVING CAR.	*Front.*
OVER THE EVER-RESTLESS SURFACE	19
IN THE SAME LATITUDE	35
OCEAN CURRENTS AND TRADE-WINDS	39
ERUPTION OF MT. EREBUS	57
"IT IS BY MEANS OF SLEDGES, AND NOT OF SHIPS"	75
SHIP WITH ALL SAILS SET.	80
3-MAST-SCHOONER, TOW-BOAT, BRIG, ETC.	83
SLOOP	86
CAT-BOAT, FISHING SCHOONER	96
THE BATTLE OF TRAFALGAR	113
A ROBBER SHIP	123
TAKING PILOT IN ROUGH WEATHER; AND BOSTON HARBOR.	141
GATHERING SEA-WEED	175
LOBSTER-FISHING; PEARL-OYSTER-FISHING	191
COD-FISHING	211

OLD OCEAN.

I. — 146,000,000 SQUARE MILES OF WATER.

OFTEN things familiar to us present an entirely different picture if only we change our point of view. I used to know a boy who, whenever he had gazed for a time at a bit of landscape, would turn his back upon it, bend over, and try how it appeared seen from between his feet. The effect was certainly very novel. Suppose, now, we take a globe, or a map of the world, and, neglecting the continents, whose outlines are well remembered, study the area occupied by water. I warrant you a new impression of our globe will be forced upon us. Very likely we never before have really appreciated the vastness of this vast, blank, intervening space between the lands. Spread out a map of the whole world, stand back and look at it. The largest continents become simply islands; the

7

great islands dwindle to islets, and the little ones
disappear altogether —

> ".   .   .   and, poured round all,
> Old ocean's gray and melancholy waste."

Looking at the land, we divide the surface of the
earth into eastern and western hemispheres ; but looking at the water, we make an opposite classification.
Encircle the globe in your library with a rubber band,
so that it cuts across South America from about
Porto Allegre to Lima on one side, and through
southern Siam and the northernmost of the Phil-
ippine islands on the other, and you make hemispheres, the northern of which (with London at its
centre) contains almost all the land of the globe,
while the southern is almost entirely water, Australia
and Patagonia being the only lands of consequence
in its whole area.   Observing the map in this way,
noticing that, besides nearly a complete half-world of
water south of your rubber equator, much of the
northern hemisphere also is afloat, you are willing to
believe my assertion that there is almost three times
as much of the outside of the earth hidden under the

waves as appears above them — one hundred and forty-six million square miles of endlessly restless surface.

But in bulk, as well as in space, the ocean is mightier than the land. There are very few mountain peaks whose topmost pinnacles tower as high above the tide-level as incredibly wide areas of ocean-bottom are sunk below it; and, while the great majority of land everywhere is only a few hundred feet above the surface of the sea, the *average* depth of the Atlantic is over ten thousand feet, and that of the Pacific and Indian oceans even greater. If, then, Nature were to plane down the earth with its mountain-ranges, in order to fill the ocean-valleys and make a perfectly smooth surface all over the globe, she would find it needful to dig away all the dry land of the globe, and also much which is submerged; and then salt water would cover everything with a uniform depth of more than a mile — just as I think it used to. That was long ago. Though we speak of ancient rocks and everlasting rivers and hoary mountains, the ocean is older than any of them — older than anything else, except, perhaps, the atmosphere, which is the ocean's

mother. Many scientific men believe that when our
planet first went circling swiftly in its orbit it was a
glowing mass of molten, mixed-up metals, minerals
and gases, which only held its shape because it was
spinning so rapidly and racing on with such speed
that no one particle of it had time to get the start of any
other. They believe it was enveloped in thick masses
of fiery vapors, and if there was any solidity about it
anywhere, it certainly was not near the surface. But
as time went on, the icy chill of space cooled these
vapors slowly down; and all the time chemical
changes went on within their masses, bringing this
and that element together and separating others;
causing some gaseous materials to make their way
towards the centre and others to seek the outside,
until, finally, *water* came into existence.

After that — let us picture it — what deluges of
rain were poured out of and down through those
murky clouds where thunders bellowed and light-
nings warred! At first, all the rains that fell were
turned to steam again, as would be a cupful of
water thrown into a blast furnace ; but by and by
the steady down-pour cooled the shaping globe so

that all the water was not evaporated, but some
stayed where it fell, and this increased in amount
more and more, until finally, between the hissing
core of the half-hardened planet and the dense clouds
which kept out all the sunlight, there rolled the
heated waves of the first ocean — an ocean not only
shoreless from pole to pole, but boiling hot, and
sending up ceaseless volumes of steam.

Yet all the while the cooling of the planet went
slowly on, and presently a crust or skin like the leather
covering of a ball formed over the hitherto almost fluid
surface of the mass.   Now when any heated substance
cools, it contracts.   The world, being of huge size,
contracted on a large scale ; and as the cooling went
on faster in some parts than in others, and unequally
at various seasons, and was disturbed by explosions
and swelling from underneath, the contraction was
highly irregular, and produced great cracks and ridges.
These were most prominent around the North Pole,
as is shown by the fact that the land of the globe is
mainly grouped around that pole, and their range
was in general north and south lines.   Such were
the results of contraction upon this first weak crust

of the earth; those parts which were stiffest resisting contraction, or simply bulging up, while great areas of thinner crust sank inward. Into these huge depressions which were constantly changing, but with less and less frequency, as the warping heat continually decreased, poured the wide waste of waters, the ridges between forming the earliest shores of this black primeval sea.

How different from these beliefs of scientific men are the fanciful notions about the ocean in the writings that have come down from the days of those great empires in Egypt and Arabia and Syria; from the kingdom of Phœnicia, the elegant civilization of Greece and the battle times of the Romans. The old Greeks, for example, regarded the earth as a flat space having a circular border, around which there was perpetually flowing a river that had no visible shore. This river was the source of all the rivers and other waters: out of it the sun lifted itself in the morning, to sink into it again on the other side of the plain; the stars, too, rose and set in its flood; and on its further banks, which no one had ever been able to reach, were the abodes of the dead. They also deified this outer water under the name of Oceanus; and when men had travelled a little further and learned that west of Europe and south of Africa there really was a boundless space of water, it was natural they should call it by the old poetic name, Oceanus — the ocean.

The ocean which the Phœnician and the Latin navigators knew almost exclusively — though they must have been somewhat acquainted with the Arabian and African edge of the Indian sea — was that west of Spain and separated from the Mediterranean by the Pillars of Hercules. To this water they gave a special name, the origin of which is explained by another myth.

One of the very oldest and most venerable of the Greek gods was Atlas. The poets told all sorts of stories about him; but whatever else he did, all agreed that he supported on his shoulders the pillars that upheld the sky. These pillars seemed to rest in the western waters, just beyond the sunset horizon. Later on, it was believed among the Greeks that out there, beyond the Pillars of Hercules, lay an island of great extent, yet forming only a passage-

way to a vaster continent beyond. Naturally, this
unseen island was spoken of as the home of Atlas,
— Atlantis, or Atlantica.

But let us return to the early offices of the ocean
waters upon this globe of ours. It is certain that
the great waters and great lands have remained
substantially in their present shape ever since they
first became distinct, the first cooling throwing up
certain portions which formed a model for the continents, leaving all the rest of the globe in a series
of great basins holding water — basins which grew
deeper by the steady contraction and sinking of the
crust in these the weakest spots.

The moment it had shores to beat upon, that
moment the ocean began to knock them to pieces
and grind the fragments under its pounding surf.
These fragments sank wherever the currents bore
them or the water was still enough to allow them
to be deposited — sometimes in coarse particles,
sometimes in fine silt — and there hardened gradually
into stone, which was arranged in layers or *strata*,
and hence is called stratified rock. The primitive
or original rock — New England's granite is as good

an example of it as you can find — contains in a
mixed state all the components of the future stratified rocks. The original rocks held the different forms
of lime, magnesia, etc., to make the limestones; the
silica to make the gritty sandstones and quartzes;
the alumina to make the clays, and so on. The sea
not only was the agent to eat this old rich crust
to pieces and respread it into strata, but to sort out
for us the materials to a considerable extent, laying
down beds of limestone by themselves, and sandstone,
shales, marl, etc., by themselves.

All this work upon a chaotic world was not done
in one way, however. While with sledge-hammer
blows old ocean's ponderous waves were crushing
exposed points of the coasts, other of its agencies
were busy elsewhere. For instance, when dry air
comes in contact with water, it soaks up as much as
it can carry in that invisible form — *moisture*. But
this dampened air, being light, is continually rising,
while the heavier dry air crowds underneath to rise
in turn with its load of moisture until it reaches a
height where the cold is great enough to turn the
moisture into drops; then it forms clouds, and by

and by these become too heavy and fall as rain.
Doubtless you know this process very well already,
but I repeat it to make sure of my next point. The
atmosphere even now contains a small proportion of
a powerful gas known as *carbonic acid;* in those early
ages there must have been so much of it that no
living thing could have breathed. Now this gas is
absorbed by water, and every particle of moisture
which the ocean sent up took along a little carbonic
acid with it, and so helped to purify the atmosphere
and make it fit for plants and animals later on.

But don't think now that the carbonic acid was at
that time all nuisance : it had its place in the plan.
This acid is almost the only thing that will dissolve
certain hard minerals, among them silica, the substance of pure sand, and the various sorts of quartz
and gems, and so enable water to take them up
and carry them along. The rain, therefore, well armed
with the carbonic acid of the thick air, by drop, by
rivulet, and by flood, was wearing down the rugged
mountains, filling up the gloomy chasms, levelling off
the surface of the rough continents, and helping the
ocean in sorting out the materials which, later on,

should become useful to man as fertile soil, or be sought after as building-stones, as metals for manufacturing purposes, or as minerals precious for their beauty.

Meanwhile, as time went on, the crust of the earth cooled and stiffened more and more until it became so stable that only slight and gradual changes of level took place outside of strictly volcanic regions.

Now I have run the risk of being voted tedious with this bit of geology and chemistry, in order to show how important a part the ocean has played, and is still performing, in remodelling the world; and also to show that it was done chiefly along the margins, building continents and large islands out at the edges just as fast as they were levelled down over their interiors ; so that, from being enormously lofty, narrow and precipitous, they became broad and (for the most part) flat. When you know that the whole vast plains of Hindostan are believed to have been added to primeval Asia, and the broad empires of Brazil and Patagonia to the central stem of South America, think what mighty mountains, to which the present Himalayas and Andes are mere hills, have been cut down to supply materials!

Yet out toward the middle of the sea, hardly even a short hundred miles from the shore, little drifting soil could be carried before sinking, and so we feel sure that no new bottom of consequence has ever been thus formed. The great basins of the sea — what vast valleys, thousands of miles wide and from one to fifty thousand feet deep, they would make if we could see them emptied! — have hardly been altered since creation.

But the ocean, like other architects, did not work for nothing; it took its pay for this big job of hewing a world into good shape in a sort of toll of minerals, which it holds suspended in its flood. The chief of these is salt, which has been dissolved out of the shores and bottom as time went on. Analyze sea-water, and you find that, besides the large amount of common salt (known to chemists as chloride of sodium), various other compounds of salt, lime, and other minerals, classified as sulphates, carbonates, iodides, bromides, and so forth, have been extracted. The lovely greens, purples, crimsons and scarlets painting

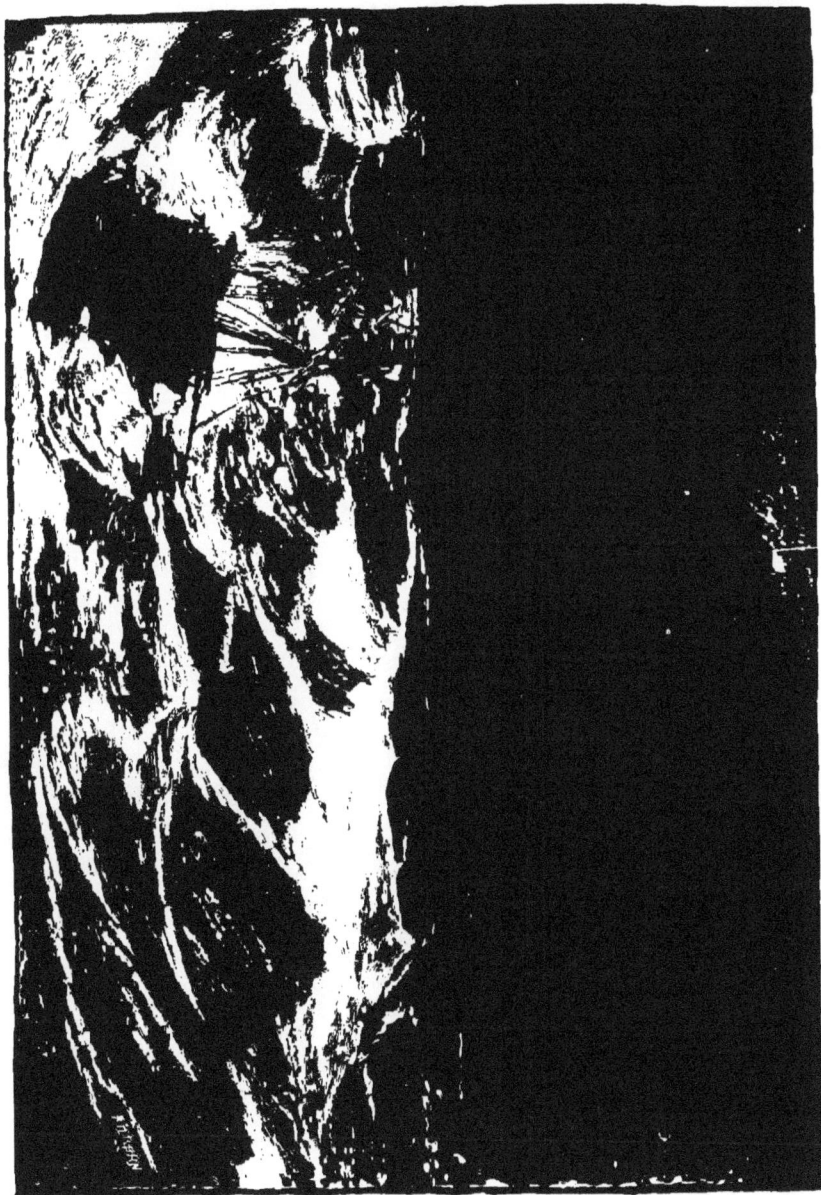

the wonderful coralines and sea-weeds dragged to light from depths of perpetual gloom, are largely dyed by these compounds.

All these salts together amount to about four parts (by measure) in every hundred of sea-water, and make about one-thirtieth of the weight of the ocean.

In many parts of the world, all the salt used is obtained by boiling down sea-water in vats; the salt being left in white crystals in the bottom of the kettle when the water is evaporated away.

It was by this means that our great-grandfathers in New England provided themselves with salt; and at certain places on the coast of Maine and New Jersey the practice has been kept up until very recently by fishermen who use a large quantity of salt in curing their fish. These old salt-boiling camps on the desolate beaches were very picturesque affairs, but have about all disappeared now; but they would come back if occasion called. An example of this happened during the Rebellion. There are no salt-wells or mines in the Southern States, and the Confederates were obliged to open factories on the coast, getting all their salt by boiling down the water, after the custom of a hundred years before.

Besides this salt, which makes ocean water much more dense and heavy than fresh water, so an object may float in the former that would sink in the latter, and swimming much easier in the ocean than in a lake, sea-water contains two very valuable metals—gold and silver; and in such quantities that it is profitable to scrape the old copper sheathing of ship bottoms to secure the silver which has formed a film over its surface. Think of silver-plated frigates!—that is what they all are after a long voyage; and somebody has estimated that the whole ocean holds no less than two millions of tons of the shining metal. Silver is now worth about $1.20 an ounce: it will be a pretty problem for you to see how rich old Neptune would be if he had it all in coins. Then to it you must add his riches in gold, which can be ascertained when I tell you simply that from each ton of sea-water one grain of gold may be extracted, and that the total bulk of salt water on the globe is said to be 290,000,000 cubic miles. These are *average* figures, of course, for the density—

that is, the amount of salt, etc., in a given quantity of
water — varies in different parts.   Deep water is salter than shallow, because the saline matters sink;
equatorial districts salter than arctic, because of
more rapid evaporation there; and the Mediterranean saltest of all, because, I suppose, it is not only
in a hot latitude, but so confined that its waters
do not change as freely as they do outside.   There
are springs of mineral water, and even of fresh water,
in certain parts of the ocean ; and some of the latter bubble up so forcibly that all the salt water is
pushed aside, and ships are said to fill their water-
casks at these sweet fountains in the midst of the
open Pacific.

The names of the different parts of the one great
ocean are familar enough, but their boundaries must
necessarily be ill-defined.   The Atlantic lies between
the Americas on the west, and Europe and Africa on
the east.   The Pacific spreads from the Americas
on the east to Asia on the west; but its great
southward expansion among the islands east of
Australia is called the South Sea, and the expansion
northward between Australia and Sumatra on the

east and Africa on the west, takes the name of the
Indian Ocean — the smallest of all.   Besides this,
we speak of all the south-polar waters as the Antarctic sea, separating them from the southern extension of the Pacific on one side, and the southern
Atlantic on the other, by the Antarctic Circle.   Similarly, the imaginary line of the Arctic Circle incloses
the Arctic Ocean.   But all these names and distinc-
tions are for convenience, and in fact there is but
one ocean, whose waters are alike and inseparable,
and always intermingling, as will be explained in my
chapter on Currents.

———

NOTE. — An account of the fabled lost island of Atlantis may be found in
Irving's Columbus, appendix to last volume.

II.—WAVES AND CURRENTS.

IF I start with the remark that the earth rolls from west to east, it is not because I think you do not know that, but to lay the foundation for something further. Surrounding the globe is the great envelope of the atmosphere, forty miles or so thick, which revolves and travels with it, and is really a part of the planet. But this atmosphere is so light that it does not revolve quite as fast as the solid earth whirls within it. It hangs back a little. If (as is near the truth) a certain point on the equator — say the city of Quito, in Ecuador — is moving at the rate of a thousand miles an hour with the spinning of the globe on its axis, the atmosphere there only moves at the rate of 995 miles an hour. Of course, such a lagging back of the atmosphere must produce a steady wind in an opposite direction to the revolution of the earth,

or from east to west, and at the rate of five miles an
hour; and such a wind actually does blow all the
year round in the neighborhood of the equator. It is
called the *trade-wind;* and it has important influences
beyond the cooling of the hot regions, or the fact that
a ship may set its sails and ride before it for weeks
without the trouble of moving a spar or losing a moment's time.

But before describing these influences upon the
great currents of the ocean, let me call your attention
to another matter — that of waves.

This simple matter of waves has caused a deal of
discussion among philosophic men who were trying to
explain it. It has generally been said that in waves
the water does not move along, but simply rises and
falls. Complicated instruments have been made to
show how this is, but you can illustrate it yourself by
fastening a long strap at both ends, not drawing it
very taut, and then giving one end a sudden jar; you
will see a line of waves run the whole length of the
strap, though certainly the leather does not move forward. But water is not confined at the ends, and it
is only a very peculiar jarring disturbance in a con-

fined space that causes water to toss up and down
without any motion ahead. On the contrary, though,
to a certain extent all waves are tremors which run
across the surface faster than the heavy water can
travel, and so a floating plank will seem to toss up
and down in the same place while the waves go rushing under it, letting it rise and fall on their heaving
ridges, yet the water in the waves goes forward all
the time; and it moves faster than the floating plank,
because it is lighter and is more easily pushed along
by the wind. Watch the sea or lake when a breeze
is ruffling its face or a gale is crowding down upon it
and urging it into violent disturbance. You see the
long curving ridges of water coming swiftly on, one
close behind the other, all marshalled in the same
direction like long ranks of soldiers, and at a thousand points breaking into lines of hissing foam. Each
of those ranks of angry green water under this gale
from the north has a long rounded curve on its hinder
slope, in the direction from which the gale comes, but
its southern front is almost straight up and down —
a wall of water, as sailors say of the huge waves of
mid-ocean, which, when you are in front of them,

seem to tower overhead as though the whole mass
would the next minute topple over upon the ship and
you.   Sometimes the crest, urged by the gale, *does*
lose its balance and crash forward in that sparkling,
resounding cataract which we call a "white cap."
Surely this water is moving ahead — racing on like a
frightened herd of white horses whose snowy manes
of cold sharp spray dash fiercely in our faces as we
sit on the open deck or stand at the end of the long
pier.   Such a tempest will drive the water upon the
coast it beats against, until it raises the level of all the
river-mouths, fills up all the bays, floods the salt
marshes and throws the surf far beyond its ordinary
mark.   On the other hand, where a strong gale blows
off a coast, it carries the water out until wide areas of
rarely seen bottom are exposed along the shore.

I conclude, then, that waves do not simply lift the
water up and let it down again, but that they also
bear it along with a speed proportionate to the power
behind them.

Now in the tropics, as I have said, the trade-wind
blows steadily from east to west right around the
earth without stopping.   This gives it a tremendous

space of water to pass over, nearly 10,000 miles in a
stretch at one place across from South America to
Australia. If one short breeze or gale can set the
sea flowing strongly upon a certain coast, and even
turn back the currents of large streams, then surely
this strong and ceaseless trade-wind will set the
water of the tropics flowing steadily in its own direction from east to west. And it does do this; but it is
greatly helped by another agent of nature.

The drops which make up a body of water are the
most restless things in the world: they are always
sliding down the least slope, sinking out of the way of
lighter substances, rising to let a heavier object pass
down beneath them, or moving hither and thither in
endless search of that levelness and quiet which we
call *equilibrium*. Furthermore, when water is heated
it becomes lighter. If, therefore, a portion of the
sea grows more warm than the rest, it will rise to the
surface; and whenever a portion becomes cooled
below the ordinary temperature it sinks.

Now under the blazing sun of the torrid zone the
surface-water of the sea gets very warm indeed and
never has any chance to cool, while in the arctic and

antarctic regions the ocean is always chilled by permanent or floating ice until it is nearly cold enough to freeze; but these masses of warm and cold water cannot remain separate in the one great ocean. The hot tropical flood continually rising *must* flow away somewhere to find its level, and it can flow nowhere except towards the poles, for there the ever-sinking volume of chilled and therefore heavier water sucks it in to take its place, while it, in turn, flows underneath toward the equator, there to fill the gap which the escaping warm water leaves behind. So we know there is constantly going on an interchange of water —a constant flowing away from the equator northward and southward on the surface, and a flowing in towards the equator along the bottom; an endless springing up in the torrid zone and a steady settling down of the polar seas. One out of many proofs of this fact is, that the mid-ocean abysses, five or ten or more miles deep, are known to be ice-cold. This could not happen unless they were constantly filled and refilled with new water from the great coolers at the poles; for if the water at those depths should remain unchanged it would quickly become very warm from the heat of the interior of the earth. But while this invisible *vertical circulation* is going on, another more visible and interesting set of movements is in progress on the surface, forming what are known as *ocean currents*. These are vast rivers in the ocean flowing across its face in certain directions and to a certain depth, as rivers make their way along the land. They begin and are kept going by a union of the two causes already explained — heat and wind.

The heat of the sun at the equator warming, lightening and evaporating the water, constantly tends to draw the colder water from the poles, and particularly from the great antarctic sea. The cold water, hastening to the equator, is soon interrupted by the extremities of Australia, Africa and South America, and so split into three great branches. That which passes into the South Atlantic goes on northward along the western coast of Africa, getting so warm under the hot sun of these low latitudes that it will not sink, as has the great mass of the water which first left with it the ice-zone, but comes more and more to the surface, until it strikes against the great shoulder of Guinea and is turned sharply westward. Now it is

squarely under the trade-wind and headed the same
way; constantly urged forward by this moderate but
endless tugging of the wind upon its waves, the current can never swerve, and flows along the equator,
and for half a dozen degrees each side of it, straight
across the Atlantic. South America, however stands
in its path, and the wedge-like coast of Brazil, pointed
with Cape St. Roque, splits this great river. Part of
it now turns southward and swings back across
towards Africa, making an eddy a couple of thousand
miles wide in the South Atlantic, and another arm
runs down the Patagonian coast. But by far the
largest part of the divided current is sent northward,
past the Amazon and the Orinoco and all that low
steaming coast of upper Brazil, in through the mazes
of the Antilles, to the pocket of the Carribean Sea,
and thence struggles out between the larger islands
of the West Indies into the North Atlantic. It used
to be thought that out of the Carribean Sea all this
moving flood poured into the Gulf of Mexico, and
thence out; but now it is known that the trough between Yucatan and Cuba is not deep enough to give
it passage, nor, if it were, is the pass between Florida

and Cuba large enough to let it out.   Nevertheless it keeps the old name, and, I suspect, will always be known as the *Gulf Stream.*

When it has worried through the islands and has once more gathered its full force together, the Gulf Stream flows northward close along the coast of our Southern States until Cape Hatteras gives it a swerve away, when it strikes out to sea and pushes straight across to Spain, where a branch leaves it and runs northward between Iceland and the British Islands, while the main body turns southward to mingle again with the equatorial current from Africa and repeat its journey all over again.   It is in the heart of this great circle of currents in the middle of the Atlantic that navigators find that dreaded region of heat and calms which they call the Doldrums; and here, too, float round and round the wide, buoyant meadows of the Sargasso Sea.

Meanwhile another most important cold stream is making its way through the Atlantic, known as the Arctic current.   It comes down out of Baffin's Bay, joins a similar flood from the outer coast of Greenland, is thrown up to the surface by the Banks of Newfoundland, and meeting with the warm air, produces those thick and prolonged fogs so common in that region, fills Massachusetts Bay with chilly water, and finally, meeting the Gulf Stream off the Virginia coast, dips under it amid that commotion of waters which makes Cape Hatteras a centre of storms.

The Pacific has a very similar arrangement of currents. The trade-wind drives westward the waters from the western coast of South America, pouring a branch down between New Guinea and Borneo, while a larger branch bends northward along the coast of Siberia, swings across to the coast of Alaska, and then on down to California, where it is gradually swerved westward on its old equatorial track. This is a warm river like the Gulf Stream, and is usually called the great Japan current. But down through Behring's Strait, which is too shallow to admit a large one, comes a small cold stream, which answers to the Arctic current of the other side.

In a lesser way, the Indian Ocean has a strong stream flowing directly across from Australia to South Africa. It is of immense help to the ships returning from China and the East Indies. There are also

IN THE SAME LATITUDE.

various minor currents, like that one south of Australia, and the one that forms a great eddy in the Arabian Sea. You will find all of them marked on a map in most geographies, and will understand me better after you have looked them up.

Not all are as well marked as the Gulf Stream. Its brightly blue water is in such contrast to the darker, greenish hue of the remainder of the ocean that sailors can often tell when they enter the edge of the current, half their vessel being in and half out of the stream. If you approach from the east you find that the water at first shows only a warmth of fifty or sixty degrees near the surface; but as you sail on, this gets higher, until, opposite Sandy Hook, you may get as high a reading on the thermometer as 80 degrees, and opposite Florida above one hundred degrees. This difference in temperature between the eastern and western margins of the Gulf Stream is owing to the presence of the great river of arctic water flowing in an opposite direction between the Gulf Stream and the shore. Off Florida the Gulf Stream is about sixty miles wide; off New York it is over one hundred miles in width, but is less sharply

defined.   Its depth is hard to determine, but it certainly amounts to several hundred feet.   It is worth
remembering that, although some guesses had been
made at it before, Dr. Benjamin Franklin was the
first man really to study the Gulf Stream and tell us
anything of its origin and course.

These ocean currents have a great influence upon
the climate of both the land and the seas, and affect
the inhabitants of both in various ways.   The study
of this influence brings out some entertaining facts.
For example, the North Atlantic is the stormiest of
all oceans, because the Gulf Stream heats portions
of the atmosphere and thus sets winds a-blowing.

Scotland lies as far north as Labrador, and the latitude of London is above that of Lake Superior.   Yet
they have none of the terrible frosts and heavy snows
which prevail in Canada and make Labrador a land
of ice almost uninhabitable.   This difference is due
almost entirely to the fact that the Gulf Stream pours
its warm flood against the coast of Great Britain, and
even tempers the Norwegian coast, so that forests
grow and the Laps can live in much comfort on a
line with the endless glaciers and frozen seas of

Greenland.   But instead of having all the sea-breezes
warmed by flowing over water that brings with it the
heat of its long wanderings under the fierce equator, the unfortunate coasts of Greenland are bathed

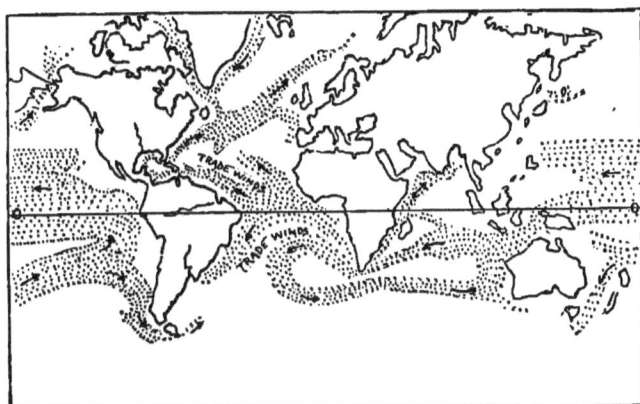

OCEAN CURRENTS AND TRADE-WINDS.

in water chilled by months of captivity near the pole,
and loaded with ice which cools down all the winds
that blow ashore till they freeze everything they
touch for half the year, and make it foggy or chilly
the rest of the time.   Hence Boston is a city of frost
and snow all winter, when it is really no further
north than sunny, flower-growing Italy, where one
laughs at winter.

Similarly, in the Pacific Ocean, the northward movement of the great Japanese current makes the coast of China habitable and pleasant clear to the sea of Okotzk, and gives the Aleutian Archipelago a pretty decent climate ; but by the time the current has had a touch of Behring's Strait and swept down the shores of British America, it has got well rid of its warmth, and gives to the Pacific coast of the United States that constant succession of chill winds and fogs and the heavy rains or snows which mar the climate of California. The weather in the interior of continents is pretty much alike on similar latitudes the world round, varying with height ; but the climate of all sea-coasts is good or bad as a place to live, in accordance with the temperature of the water which the currents bring to that part of the ocean.

But the currents of the ocean influence something besides the weather. Upon them depends to a considerable extent whether a certain part of the coast shall have one or another kind of animals dwelling in the salt water. This is not so much true of fishes as it is of the mollusks or ' shell-fish," the worms that live in the mud of the ˉtide-flats, the anemones, sea-urchins, starfish and little clinging people of the wet rocks, and the jelly-fishes, great and small, that swim about in the open sea.

Nothing would injure most of these "small fry" more than a change in the water making it a few degrees colder or warmer than they were accustomed to. Since the constant circulation of the currents keeps the ocean water in all its parts almost precisely of the same density, and food seems about as likely to abound in one district as another, naturalists have concluded that it is temperature which decides the extent of coast or of sea-area where any one kind of invertebrate animal will be found; for beyond, the too great heat, or else the chill of the water, sets a wall as impassable as if of rock. It thus happens that the small life of hot Cuban waters is different from that of our Carolina coast; and that, again, largely separate from what you will see off New York; while Cape Cod seems to run out as a partition between the shore-life south of it, and a very different set of shells, sand-worms, and so forth, to the northward. This is not strictly defined; many species lap over, and a few are to be found the whole length of our coast; yet Cape Hatteras ends the

northern range of many half-tropical species, and
Cape Cod will not let pass it dozens of kinds of animals abundant from Massachusetts Bay northward.

But out in the ocean, the warm current of the Gulf
Stream forms a genial pathway along which southern
swimming animals — like the wondrously beautiful
Portuguese man-o'-war — may wander northward for
hundreds of miles beyond where they are found near
shore ; but if by chance they stray outside the limits
of the warm Gulf Stream they will at once be chilled
to death. Similarly the arctic currents let arctic animals, used to half-freezing water, make their way as
far south as Massachusetts Bay.

There is another thing of interest about ocean currents. They not only allow swimming animals to go
beyond their ordinary range by supplying them with
water of the right temperature, but they carry floating
burdens where they are greatly needed. They bring
the icebergs — though perhaps these are not among
the things needed, since they help to form those
terrible fogs of Alaska and Newfoundland ; and they
often bear the great logs which come floating down the
Amazon or our own rivers clear across to the shores

of Europe. Before the western half of the world had been discovered to Southern Europe by Columbus, these water-soaked, weed-grown, barnacle-flecked trunks of unknown trees used to puzzle men over there greatly; and the conviction gradually forced itself upon their minds that there must be an unseen country far away to the west where these trees grew. Thus the Atlantic currents bore messages from the mysterious new world which finally were heeded by brave explorers. They gave a hint of the way to America just as the buffalo-trails first taught the engineers the best routes of our Pacific railways across the Rocky Mountains.

The currents do another service to the world. Where they strike islands not far from some continent or some other islands, they often carry along old logs with plants growing upon them and quantities of seeds which are not injured by their short voyages. When, therefore, the coral polyps build up one of their reef-islands until it appears above the waves, thither the currents bring roots and seeds from neighboring islands, and quickly plant them upon the new barren shores, so that in a few seasons the little islet becomes green and wooded and ready to hold its own against the winds and waves.

Moreover, the same drifting stuff will carry many sorts of land animals, — insects, snails of many kinds, reptiles, and even four-footed beasts, and so not only give the island a vegetation, but populate it with many of its smaller animals. This seems to you, perhaps, a very accidental and haphazard way of fitting out an island so that presently it may support human beings, nor is it the only means by which barren islands become productive ; but it is important as far as it goes ; and when we study into the distribution of plants and animals in an archipelago, we are pretty sure to find those of the same sort upon islands that lie in the same current.

III.— EARLY VOYAGES, AND THE SOUTH POLE.

I SUPPOSE that by the discovery of how to get a fire, the first savage men were able to make the longest of all the steps toward civilization. Next to this, the tilling of the soil has been of most advantage; but surely the third means of growth out of barbarism has been the knowledge of navigation. As we have no history of a time when men did not possess a fire, so we cannot go back to where they did not have boats. Yet it is easy to return to an age when all use of boats was confined to inland waters, or to ocean coasts where headlands were always in sight as landmarks. As wars and trade called for larger enterprises, and knowledge grew, the boatmen became more venturesome, until, finally, a wonderful invention enabled sailors to leave the land behind, and become real mariners, pursuing a steadily straight course for weeks together across the untracked sea.

How well the early Chinese knew the oriental seas, and that the Polynesians could steer from island to island for thousands of miles, we can only guess; for the first seafaring people of whom we have any accurate history are the Phœnicians, who were at the head of the world about 500 years B. C. Their capital city was Tyre, on the Syrian coast; and when we first hear of them, their ships traded east and west from England to India. As long ago as 600 years B. C., Necho, a wise king of Egypt, employed Phœnician sailors to explore the eastern coast of Africa, and they started down the Red sea. Two years later they came sailing into the Mediterranean through the straits of Gibraltar, showing that they had rounded the Cape of Good Hope. A century or so later the princes of the second great Phœnician city, Carthage (now Tunis), sent Hamio, with sixty ships, to explore the Atlantic coast of Africa; but he seems to have got only about as far as Sierra Leone. Then in 320 B. C., an expedition from Massilia (where Marseilles in France now stands) set out northward under another Carthaginian captain, Pytheas.   It sailed along the coasts of Spain and France, clear around Great Britain, and back home, and is famous for its discovery of Ultima Thule (now generally thought to be Shetland, or else a part of Norway), then considered the northern extremity of the earth.   The next year Pytheas penetrated the Baltic.

Meanwhile, exploration in the east had been pushed by the conquests of Alexander the Great, but chiefly on land, so that all the coasts beyond, nearer India, remained unknown for hundreds of years later; though the fact that the earth was a globe was understood two centuries before Christ; and the first geographer, Eratosthenes, suggested that by sailing westward new continents and islands might be discovered beyond the "utmost purple rim" of the Atlantic.   This is about all that the famous Ptolemy knew in the second century after Christ.

The conquests of Rome, Constantinople, and the Mohammedans, extended a knowledge of Asia; but in Europe more than six hundred years went by after

Ptolemy, before any progress was made. This was
that dark time called the Middle Ages, which followed the weakening and breaking up of the Roman
empire. The first signs of its brightening came
from Gothland, on the cold shores of the North sea,
where the Norman Vikings ruled, not only the most
enterprising and refined race in western Europe, but
all-powerful on the sea. Dreaded as these rough
old Norman sea-kings were, they were not wholly
pirates and marauders; and from them the proudest
Englishmen and Americans trace their descent.
Well, these Normans, with their friends the Norse-
men, braving the icy terrors of the wild seas in their
small ships, by the middle of the ninth century had
sailed past North Cape, and all along the coast of
Lapland, knew all about Scotland and the Scottish
Isles, had discovered the Faroe Islands (A. D. 861)
and settled them, and in 874 planted colonies in
Iceland. Thence old Red Eric — who has atoned
for his crimes by his valor— bore away westward,
until the bleak cliffs of Greenland loomed before
him, and landed upon the continent of America
five hundred years before Columbus started from

Cartegena. The colonies thus established in Greenland not only kept up communication with Iceland
and Europe, but made new explorations for themselves. They sailed around into Baffin's bay, going
well towards its northern extremity; and before the
year 1000, Byorni, a son of one of old Red Eric's
comrades, led an expedition southward past Labrador,
around Nova Scotia, and away south of Cape Cod to
a pleasant district called Vinland, where a Norse
village existed for many years. Just where this was
situated we don't know, but it is thought that the old
tower at Newport, Rhode Island, is a relic of those
early days. Very soon after, fishermen began coming
from Norway, Jutland (Denmark), and the Briton
coast of France, to the banks of Newfoundland;
and it is a curious fact, that the Indian name in
Newfoundland for the cod is not an Indian word at
all, but a corrupted French word.

For five hundred years these Norman colonies
flourished, and the Newfoundland banks were annually fished upon by Europeans, who found their way
back and forth by the help of the stars, I suppose.
But toward the end of the fourteenth century catas-

trophe came. Fleeing from destruction, a Tartan
chief with a few followers marched to Kamtchatka,
crossed Behring's straits after much fighting, collected an army of Arctic Indians, and made his way
clear across from Alaska to Labrador, where he arrived in 1399. There he heard of the Norman towns
in southern Greenland, rich in things he needed.
Not minding the wrong of it, he built boats enough
to carry a crowd of his savages across, attacked the
villages then feeble with the devastation of a plague,
killed all the people and left the towns in ruins. This
was the end of all colonies in Greenland ; for just
then their people at home were too busy in conquer-
ing England and fighting the Gauls to think of America, and the far western shore was forgotten by all
but a few students.

But about this time the world was waking from its
long stupidity. Commerce was reviving, learning
began to flourish, and kings were willing to let their
subjects plan and carry out naval enterprises. Just
in this ripe time came the help which enabled navigation to take the greatest step forward it ever did or
ever will take. It appears that while the western

world was under the cloud of the Dark Ages, the
Chinese discovered — no one knows just how or
where — that a bit of iron properly magnetized will
invariably turn so as to point toward the north, no
matter in what part of the world the test is made.
Europe learned of this astonishing invention first at
Naples about 1307, and the value of it was seen at
once, for it gave the sailor a sure indication of where
north lay when he was out of sight of land and all
the stars were hidden in storm-clouds. Knowing
where north was, he could easily find east or west ;
but to aid him, these, and three hundred and sixty in-
termediate points called " degrees," were marked in
a circle on a small sheet of paper, over which the
magnetized needle was so fixed as to swing freely ;
and this arrangement, so simple yet accomplishing so
much, made the *mariner's compass.*

At this time nearly all the commerce of Europe
was with India and China. The overland route was
long, expensive and dangerous. The water route was
equally so, for vessels had to stick close to land, and
thus were often on a perilous lea shore. The first
need of the world was the discovery of some straighter

and quicker road to the east. But upon this errand,
Venice took the lead, and sent Zeni to the westward
as early as 1380, but he could not get past North
America, and so returned. Then Portugal came forward under the brilliant leadership of Prince Henry
the Navigator, who, by the way, was half an Englishman, since his mother was Philippa of Lancaster. It
was Prince Henry's ambition to find the sea-path to
the east, and he enlisted the help of the best navigators of every country. Thus urged, Portugal's ships
sailed further and further southward, seeking to get
by Africa, until in 1486 Dias fought his way through
the storms that guarded the Cape of Good Hope,
and opened the route for Vasco de Gama to push
straight across the Indian ocean to Calcutta in 1497.
The sea-path to India was found at last, and the
extension of commerce to China and the great eastern
islands quickly followed; but poor Prince Henry was
dead long before.

Portugal was so much occupied with these and other
ocean surveys, and in governing her new eastern possessions, that she would not listen to the plans of an
ambitious young sailor from Genoa named Christo-

pher Columbus. He had been reading all the travels he could get hold of, and diligently studying all
that was known of geography and navigation. He had
heard of the voyages of the Vikings to Greenland and
Vinland, of the fishing trips to Newfoundland and of
the researches of Zeni. He had faith in them, and
thought if he went further south he might either get
past Vinland and sail straight on to China and eastern
India or else he would come upon an unknown continent. Either of these results would be glorious. But
it was long before he could get any government to
support the attempt. How, finally, he persuaded
Spain to fit out his small ships; how bravely he kept
on his way across the rough Atlantic with instruments
which now we should think utterly worthless, since
the quadrant and sextant were then unknown; how,
in 1492, the Bahamas were reached ; and how in sub-
sequent voyages more and more of the West Indies
were surveyed, until in 1498 he set foot upon the
mainland of South America — are already familiar to
you.

Europe was quick to profit by the discovery of the
western continent, and sent scores of expeditions to

take possession of anything not yet claimed by Spain; but the Spaniards kept in the lead, and rapidly explored not only the Brazilian coast, but crossed over and set their standards in the surf of the Pacific. Thence they extended their expeditions from Panama northward as far as Vancouver's Island, and southward to Chili.

All this time the Dutch and Portuguese were busy exploring the region about Spitzbergen and in the White sea, and possessing themselves of the East Indies, while the English and French sent voyagers to the rediscovered shores of what is now Canada and New England, until within a very few years from Columbus' voyage the whole Atlantic and much of the Pacific coast of both Americas was fairly well known.

Of all these expeditions, however, the most brilliant was that of Magellan, a native of Portugal, who commanded an expedition of three ships for the king of Spain. Steering straight for Brazil, he worked his way southward to see where the end of the continent was, and finally entered a gulf, the southern shore of which he called Terra del Fuego, because he saw so many fires there. Sailing into this gulf, he was delighted presently to emerge into a new ocean on the other side. Shaping his course northwest, at the end of three long months Magellan reached the East Indies, and knew that he had been around the world for the first time in history.

Following him with an expedition in 1577, Sir Francis Drake of England was driven by a storm west of Terra del Fuego to its southern point, and so discovered that the Atlantic and Pacific were joined there. Then he sailed northward, entered the harbor where San Francisco now stands, and then crossed the Pacific, homeward bound. Thirty-eight years later Cape Horn was rounded for the first time by the Dutch Captain Van Schonter, who named it after his native town in Holland. Meanwhile the same nation had caught sight of Australia, and in 1642 Captain Abel Jansen Tasman left Batavia on a voyage southward, which was destined to prove very important indeed, for it added to the map Southern Australia, Van Diemen's land, New Zealand, and much information concerning New Guinea and many small islands. Beyond this, the voyages of the famous English Captain Cook taught us most about the south seas where Cook finally lost his life.

During this time accident more than design had contributed some knowledge of the ocean about the south pole, although steady explorations were being conducted in the Arctic seas, to which I shall devote the whole of my next chapter.

More than two hundred and eighty years ago the existence of islands far to the southward of any continents became known to navigators, who were driven thither by bad weather, and little by little was added to the map of this desolate region ; but it was not until 1774 that any one went into that terrible Antarctic sea for the express purpose of a survey. This man was the intrepid Captain Cook, and though he went a third of the way around the globe in his efforts to find an entrance through the icy barrier, he could never penetrate beyond 71° south latitude, only about equal to North Cape, or the town of Upernavik, in the Arctic region. Later captains did little better, until 1841, when Ross, in his ships " Erebus " and "Terror," skirted the edge of the thick ice that everywhere clothed the land, though it was midsummer, and finally reached the base of the southernmost

ERUPTION OF MT. EREBUS.

land yet known on the globe — a magnificent mountain chain stretching away to the south from latitude
77° 5′. Some years before this, Captain Parry, an
Arctic explorer, gave the name of "Ross" to the
most northerly land then known; this southward
end of the world Ross now called "Parry Land,"
and so returned a compliment in a way it is not often
possible for men to do.

The most conspicuous point of all this range of
polar mountains was a lofty volcano — Mt. Erebus —
12,400 feet high, and covered with everlasting glaciers
and snow from its lonely crest to the tempestuous
water's edge. It was in active eruption, and Ross
tries to tell of the splendor of its action when the wide,
glistening waste of snow and the deep blue of the
ocean were lighted by the column of fire and smoke
hurled thousands of feet skyward from its crater;
but who can picture the grandeur of such a scene!

Meagre as this information is, it is about all we
know of the globe within the Antarctic circle, and
we are likely never to learn much more. In a latitude much further from the pole than where, in the
north, vegetation is abundant, and men and animals

live all the year round, the severity of the Antarctic climate cuts off all life, and constantly seals the water under a cap of ice. The only land appears to be volcanic; and the soil, or, rather, the structure of the islands, is often found to consist of nothing but alternate layers of ashes and ice, that have succeeded one another season after season. Most of the coast is unapproachable on account of an unbroken belt of cliffs of perpetual ice; and it is only in the outermost islands that even coarse grass, a few lichens, or simple seaweeds can be found; for the volcanic heights within are utterly destitute of any vegetation whatever, and the highest noonday heat of summer is only a little above the freezing-point.

Why this intense cold and dreadful desolation exists so much further from the pole in the southern than in the northern hemisphere, I need hardly explain to you; for you will recall that in the north the continents are so broad as to form almost an unbroken wall about the narrow polar sea, confining its cold waters, warming the air by wide radiation, and guiding the heated flood of the Gulf Stream

straight into the chilling northern sea. In the
Antarctic region, on the other hand, an immense
breadth of water is broken by no land of any
account; there is and can be no great warm current
to temper the sea-water, and along hundreds of miles
of glacial cliffs icebergs are daily breaking off and
drifting far northward to chill both water and air
beyond the limit of animal endurance. Terra del
Fuego stands as a type of all that is cold and
desolate, yet it is no nearer to its pole than the
Highlands of Scotland are to the northern "hub,"
and considerably more distant than the great city
of St. Petersburg, the capital of "all the Russias."

Of course, then, you would not expect land-
animals to be found in this vast iceland capping
the southern axis of the globe ; yet of sea mammals
there is a large variety, including several whales,
dolphins and their kin, and various sorts of seals,
small and large — notably the huge sea-elephant, now
becoming very rare. All these feed on fishes, which
are abundant there, as also are the humbler orders
of animal life. Then, too, the Antarctic islands are
the resort of enormous flocks of sea-birds : ducks,

albatrosses, penguins, petrels, etc., all different from
the arctic species. Some of these birds are giants
of their kind, as, for instance, the great "break-
bones" petrel, whose powerful beak is four and one-
half inches long. It ordinarily lives on fishes or on
floating carrion, but now and then it attacks living
birds, even those as large as a loon, killing them by
repeated blows on the head. Feeble mates of its
own species, even, are thus struck down and torn
to pieces. As for the albatross, it is only rivalled by
the condor in size and strength ; while the big,
stupid penguins seem far more fishy than bird-like,
scarcely ever visiting the land except to lay their
eggs and hatch their young.

IV.—THE POLAR REGIONS.

EVER since the sea-route from Europe to India and China was settled and the coast of South America explored, the regions within the Arctic circle have been the favorite field of discovery. It occurred to every navigator that as a way was found past the southern end of the American continent, so one around its northern border might be disclosed; and perhaps, also, a ship-route along the northern coast of Siberia and down through Behring's straits. Both of these would be far shorter than going around Cape Horn or the Cape of Good Hope. Thus captain after captain headed expeditions of research, from Barentzoon in 1596, to Mackenzie in 1789, until, at the beginning of the present century, the mainland coasts of both continents surrounding the north pole were well known. If you look at a map of the polar world you see that these coasts form a fairly good circle of land, broken only by the Atlantic ; and also that all points on this coast-line are nearly at equal distance from the pole, except at the northeastern corner of America, where a dense group of islands extends northward. Within this great circle is an unknown region about one thousand miles wide in its narrowest part, and covered with almost endless ice and snow. Whether it be all a frozen sea, or partially open water, or an archipelago of glacier-capped islands, or a mass of land, no one knows.

What hardships the men of artic expeditions have undergone, and the splendid courage they have shown, cannot be told here. You can read it in their own words, and gaze at their own hurried sketches. I know no more thrilling tales than these narratives of Beechy, Sconsby, Ross, Parry, Kane, Hayes, Hall, Markham, Young, and the " Polaris " party. And there are many keenly enjoyable arctic books in French, German and Swedish, for none of these nations have been behind us.

Well, it having been found that the dream of a " northwest passage " to China must be abandoned, and that, although Professor Nordonskjold carried his steamer through the Siberian sea from Copenhagen to Yokohama, yet this route was not really useful to merchants — what has been accomplished by three hundred years of steady and intelligent cruising in the frigid zone?

Looking first at the mainlands, we notice that they are much alike all the way round. Back from the coast lie low, level plains, forming the "barren grounds" of America and the "tundras" of the old world. These plains remain treeless, because swept during the greater part of the year by fearfully cold and fierce winds; but they are overspread with mosses and lichens so dense as to form peat-beds in damp spots. This dreary monotony is relieved in sheltered places by scanty grasses and a few flowers. Portions of this area, particularly along the lower parts of the great Russian rivers, are morasses with islands of half-firm ground to break their wide watery solitude.

In America these plains extend as far south as Labrador. West of Hudson's bay, however, the forests grow much further to the north, and continue

to encroach upon the barrens until, towards the mouth
of the Mackenzie, they reach as high as latitude 71°,
after which they fall away again to the Alaskan
coast. The same irregular boundary between the
forests and the tundras exists in Russia, the woods
growing nearest to the coast along the Lena river,
and furthest from it west of the Yenesei and in
Kamtchatka. These forests are all evergreen trees
(chiefly firs), and of diminutive size, even when of
great age, because the short summers permit only a
trifling growth.

To these desolate marshes in summer come from
the south hosts of water-birds to make their nests,
followed by hawks and owls intent upon rearing their
own young, and preying upon their weaker fellows.
Various small animals—hares, mice, lemmings, etc.
—live there; the musk-ox and reindeer leave the
insect-plagued woods to graze upon the reindeer
moss, and the fox, lynx and bear range along the
forest edges. But in winter nearly all signs of life
have disappeared, and limitless fields of snow closely
blanket lakes, rivers and grassy plains alike.

In the summer months, too, barbarous people

wander over these barrens—in Russia the Laps, the Samoyeds, the Tchooctchi, and others—and in America the Innuit and other Indians and the Eskimos; but in winter they all retreat into the forests, or else set up their warm huts on the sea-shore.

It has been said that the continuance of our world as a suitable place for us to live, depends on these vast tundas remaining low and flat as they are now. If there were mountains or even hills instead of plains, lofty cliffs in place of marshy shores, the elevation would make them so cold that the snows of winter would never melt, but harden into ice every summer, as happens on a small scale in the Swiss mountains. This would go on increasing over so wide an extent, that finally the weight of accumulated ice would disturb the balance of the world, and the result would be the destruction of animal life. As it is, we need not fear any such catastrophe.

In the Arctic seas the perils of navigation are doubled. In addition to the storms and mishaps of ordinary voyages, the seaman must here thread his way through narrow channels whose depth is imperfectly known, will often find himself wrapped in

dense fogs or blinding snow-storms, must watch lest
he be crushed in ice that is drifting, or dashed to
pieces against ice that is firm.

Sailing northward in summer, navigators first begin
to meet with great floating masses, called *bergs*, on
a line between Ireland and Newfoundland; though
occasionally bergs drift almost down to Nova Scotia.
Next they come to *drift-ice*—fragments large and
small moving in fields, loose enough to let a ship
make its way through. Beyond this is seen the
pack-ice; that is, broken ice so firmly packed together that it cannot be penetrated; and if this is of
great extent, or is a solid portion broken off the main
field that has covered the whole sea or bay, it is
spoken of as a *floe*.

The position where hard pack-ice is pretty sure to
be met with is now well known to the whaling and
sealing vessels, and the exploring ships that annually go into northern waters. It stands as a barrier
hundreds of miles wide along eastern Greenland, extends from the southern point of Spitzbergen eastward almost to Nova Zembla (which has an ice-field
of its own on the eastern shore), then stretches

across to Taimyr gulf. Thence only a narrow space
of open water is ever to be found—and that not
surely—between the solid ice and the mainland away
eastward along Siberia, across Behring's straits, and on
beyond Alaska to the mouth of the Mackenzie river.
There the great number of islands make the region
so warm, by soaking sunshine into the land, as it
were, and storing it up to be let out gradually through
the winter, that a good deal of open water is seen during the warm half of the year, extending far to the
northwest of Greenland. In the Atlantic, the fading
warmth of the last end of the Gulf Stream is sufficient
to keep the western shores of Spitzbergen pretty free
from summer ice, and enables plants to grow and
some animals to live on those far Arctic islands.

It is for this reason that the most nearly successful
of the attempts to reach the north pole have been
those by the way of Greenland or else past Spitzbergen.
These attempts have been many ; and though while I
am writing, news comes of the wrecking of the
Jeannette north of Siberia, word is also brought
that a new ship of Arctic research is to be prepared
in New York. It seems as though men would never

stop till they had sailed over the very end of the earth, and been able to write in their log-books, "no longitude, no latitude."

Do you ask, "What is the use of that, when it is so difficult and dangerous and expensive?" Well, we have learned that there is no commercial gain to be got, no people to be benefited; and we should probably discover little more than we now know of the geology and the scanty plants and animals of the unexplored regions. Astronomers say they should be glad to have certain scientific experiments made right at the pole to match with those made elsewhere in the world; but I fear the search has come to be chiefly a matter of pride. Every man who tries, knows that if he succeeds his name will be spoken with glory all over the world, and each nation is anxious that its citizens should be able to claim this grand praise. So brave sailors and scientific men will still risk their lives to go, and governments will gladly furnish them with ships.

To go on a polar exploration, a ship has to be made doubly strong by every kind of extra planking and bracing. None but steam vessels are sent nowadays, and they are usually accompanied by a second vessel which goes with them to Disco, or some other Danish settlement in Greenland, and carries an extra supply of coal and other things, so that they may start with full bins on the very edge of their field of work. After that they use their sails as much as possible, instead of steam, and so economize their coal. At the settlements of the Greenlanders they take on board fur-clothing, sledges and Eskimo dogs, and perhaps an Eskimo family as interpreters, with dried seal and walrus meat to feed the dogs. Years ago, canned meats and other provisions were not known, and a two years' trip of this kind was a long period of half starvation and sickness; but now very good food is taken in compact form, and there is always a skilful surgeon in the party. Indeed, it was only a few years ago that Lord Dufferin went to Arctic regions in his steam yacht as a pleasure trip !

It is summer when they start, and generally they can pick their way through loose ice beyond the upper end of Baffin's bay. After this begin great perils as they push northward. The channels between the many islands are narrow and tortuous, and

through them come drifting enormous bergs, towering
hundreds of feet over the mast-head, and often pitching over. It is often almost impossible to avoid
them, or the shore ice, because these island channels
cause the tides to form swift and changeable currents
which give the pilots great difficulty in steering.
Sometimes it happens that two currents will oppose
one another side by side, so as to bring two huge
bergs together, and the ship must work hard to keep
from being between when the collision comes. Then
there are the packs to be studied, which is the business of a single experienced officer called an ice-
pilot. The currents play havoc with these packs too.
One freak is to set them spinning. Let two great
whirling packs, each five hundred or a thousand
acres in extent, strike one another, and you can imagine what a crashing and grinding there will be.
The strongest ship would amount to nothing more
there than a peanut-shell under a trip-hammer.

Often ice will bar a whole channel, and from the
mast-head can be seen nothing else ahead. Then the
ship must be moored to its edge and drift with the
floe, watching it carefully to see that something dan-

gerous does not occur, until a crack opens and one can sail in. Perhaps after a few hundred yards of cautious progress this crack will close up, and then there is great danger lest the ship shall be cut in two by the pinching edges of the re-united floe, or held a prisoner and drift helplessly southward until the whole season is wasted.

These and a score of other perils avoided — and in some seasons no one can make any headway, while in other years progress is comparatively easy — the explorer finds himself at the end of the summer as far north as he is able to get, and either hastens back or goes into winter-quarters. Generally the latter plan is adopted, and a good harbor chosen; but sometimes the ship is obliged to winter wherever it happens to be.

The spot having been chosen, several anchors are set very firmly, all the "running rigging" is taken down, a wooden house, which has been brought in pieces, is fitted together over the whole deck, and a great wall of snow is built on the ice around the ship as high as her bulwarks, to keep off the wind, while some snow-huts are built near by as store-houses.

And now the days grow so short that only at mid-
day does the sun shine, and long before Christmas it
is almost as dark at noon as at midnight.   You know
how this is, don't you?

How the sun being so far north of the equator in
summer, can be seen at the pole to roll round and
round the horizon, instead of in a sideway-curve overhead ; and how, oppositely, when it has gone to the
south of the equator in winter, though equally visible
to the seals and penguins of the antartic sea, the
dwellers about the north pole get no sight of it at all
for nearly six months.   To the terrors of the bitter
cold of the winter — cold which has been known to
go as low as 60 degrees below zero and freeze ether
— are added total darkness, except for star-light.

But these winter months are not days of idle gloom
to the explorer.   Harnessing his teams of dogs to
his sledges, he pushes his way over the roughly frozen
sea, and seeks to gain a point far north of where his
ship has been able to go.   It is by means of sledges
and not of ships that the highest latitude known
to man — about 84°—has been reached; but many
solitary graves mark their tracks.

"IT IS BY MEANS OF SLEDGES, AND NOT OF SHIPS."

Too often the winter's gales and the slow crowding of the ice-floe have broken the ship to pieces, or lifted her up on to the ice and placed her so that she cannot be relaunched. Then the crew must take sledges and small boats and work their way homeward as best they can. The whole history of voyaging can tell no stories to equal the adventures of these ship-wrecked Arctic crews, and there is hardly a single record of an expedition which does not contain some narrative of the kind.

Such boat-journeys would in almost every case have failed had not the castaways been able to get birds and animals to feed upon, since they would be unable to carry away from the ship enough preserved food. No point has been reached, however, so far north that animals did not live there. Even the wild musk-ox in America and the reindeer in Europe and Asia have been seen as far north as men have seen anything; and cases are known of reindeer bearing the brand-marks of European herds, having been killed in Greenland, so that they must have come across. The polar bear roves throughout the polar regions near the coast, foxes are pretty abundant, hares occur on the uttermost islands (and the hare is a plant-feeder, remember), and in Siberia the rat-like lemming is abundant. As for the water-life, the Arctic oceans harbor whales of several species — and they are sometimes caught in Baffin's bay with Siberian lances sticking in their backs — walruses, half a dozen species of seals, which keep open holes in the ice, and make burrows beneath the snow which covers it, and so pass the winter, and a variety of fishes and shell-fish. These afford food to hosts of water-birds which retreat in winter only so far south as will give them light enough to see to fish, and in summer fly northward to nesting-places often far within the charmed circle which yet defies our exploration.

Where there is so much animal life, humanity can dwell; and we find the whole of the Arctic coast-lands haunted by scattered bands of degraded Eskimos, whose habits and history are of the greatest interest, not only because they live in such eternal desolation, but because they are believed to be the remnants of the most ancient of all the natives of our continent.

V.—SHIPS AND THEIR RIGGING.

ALONG with progress in sea explorations has gone a like progress in ship-building, and at about an equal pace. That wood would float, that a log hollowed out served better as a support in water than one whole, and that an imitation of a hollowed log, made by putting several pieces together, was best of all, were among the first facts learned by savage men. Thus came the canoes and first rude boats. For rivers and bays these proved enough, and ages passed before any advance was made.

The idea of rigging is equally simple of explanation, though some romantic stories are told by the old mythologists as to its origin. That a strong breeze moves a canoe out of its way, and that, if a man in a canoe holds a skin outstretched or a thick bush upright, the force would

79

send him along as fast as he cared to go, without the
labor of paddling, were facts quickly and gratefully
seized upon by the earliest boatmen. To have a skin
ready for the purpose, and to set up a pole to hold it

SHIP WITH ALL SAILS SET.

SAILS: 1, Outer jib. 2, Jib. 3, Foretop-mast stay-sail. 4, Fore royal.
5, Foreto'gallant sail. 6, Upper foretop-sail. 7, Lower foretop-sail.
8, Fore-sail. 9, Main to'gallant stay-sail. 10, Main top-mast stay-sail.
11, Main sky-sail. 12, Main royal. 13, Main to'gallant sail. 14, Upper
main top-sail. 15, Lower main top-sail. 16, Main-sail. 17, Main spencer.
18, Mizzen royal. 19, Mizzen to'gallant sail. 20, Upper mizzen top-sail.
21, Lower mizzen top-sail. 22, Cross jack. 23, Spanker. 24, Flying jib.
25, Foretop-mast studdin'-sail. 26, Foreto' gallant studdin'-sail.

in position, were easy matters; yet in this simple arrangement you have the first sail.

But skins were heavy and too valuable for such a purpose. People who spent much time on the water, therefore, like the ancient Egyptians, and the islanders of the Chinese and South seas, soon devised a way of weaving rushes or splints of bamboo into broad mats, and thus were able, on account of their lightness, to carry much larger and more effective sails, which were kept outstretched by one or more cross-poles or spars, and could be taken down quickly. Many such sails are in use to this day among the boatmen of Eastern nations. With the discovery of how to make cloth out of hempen, cotton, woollen and silken fibres, came a still better material for sails and ropes.

In the same way it was only gradually that men learned the best shape for the hulls of their boats. From the very first, I imagine, canoes were pointed, — at least in front; but this made the original dug-out more than ever liable to upset, and a smooth, round bottom causes a boat to drift sideways. At first the latter difficulty was avoided by putting a

board or stiffened mat over the side, so as to
counteract the leeway caused by the wind; but it was
not long before the keel was added to the bottom,
which cured both disadvantages.

It is said that the oldest forms of paddles of which
we have any record among the Egyptian or Assyrian
hieroglyphs show them to have been shaped somewhat like the hand and arm, and that similar paddles
were to be seen only a few years ago on the canals
in Holland.   This is natural, because undoubtedly
the very first paddle ever used was the naked hand.
Short paddles were soon found less powerful than
long ones; but in order to work these latter it was
necessary to brace them against something in
the middle.   Notches were therefore cut in the edge
of the boat, your paddle has become an *oar*, and by
and by the notch is to be replaced by a carefully
formed row-lock, and boatmen will learn that it is
best to feather their oars when they row.

Now build a great canoe-shaped hull, able to hold
a hundred men, out of hewn ribs and planking; set
up one or more short masts with a square sail hung
on each of them, arrange as many oars as can be

3-MAST SCHOONER. TOW-BOAT. BRIG. SLOOP. SHIP. CAT-BOAT. PILOT BOAT. OCEAN STEAMER.
HERMAPHRODITE BRIG. BARKENTINE. FISHERMAN.

accommodated, with a single long oar in the middle behind, or two oars, one on each side of the stern, to steer by, and you have the plan of the vessels that the Phœnicians and Carthaginians carried their freight and fought their battles in; that Cleopatra, in all the splendor of her court, sailed down to Alexandria with; that brought tin from Britain, that later figured in the adventures of the Northmen and formed the navies of western Europe.

So, though the trading-vessels of the ancients depended on the wind for the most part, yet they all carried and used oars; while for the ships of war or pleasure, which were made longer and narrower than the traders, sails were held to be of small service, and oars alone were employed. In the time of the Roman Empire these row-boats-of-war were so made as to allow two, three or even more tiers of oars, one over the other, to be worked by rowers who did none of the fighting and were protected by high bulwarks. Some of these boats, then and several centuries after, were built of a size almost equal to large vessels nowadays, and their oarsmen were trained into that precision of moving in perfect unison which

alone could propel such heavy craft.   Their construction was careful, too.   They were often, if not always,
sheathed with lead, and were put together with copper
nails, since iron spikes rusted out.   It was the
fashion to dye the sails in brilliant colors or to

SLOOP.

SAILS: 1, Jib.   2, Gaff top-sail.   3, Main-sail.

embroider them, and the cabins were often magnifi-
cently furnished.   Do you not remember Ezekiel's
description of the vessels of Tyre?

Their ship boards are of fir trees of Senir, their masts

*of cedars, their oars of oaks of Bashan, their benches of
ivory, their sails of fine embroidered linen.*

These row-boats held their place as war vessels
through 2000 years of history, beginning with the
battle of Salamis in 480 B. C. which saved Greece
from the grasp of Xerxes.   There were 380 open boats
on the side of the Greeks there, each carrying about 18
soldiers besides the rowers, who sat amidships (or in
the centre), while the soldiers stood upon two platforms,
one at the prow and one over the stern.   The ends
of all those ancient boats were built up into a high
projection forward, which ended in a serpent's head
or a swan's neck, the face of an owl or some odd
image; and we keep the tradition of it until now by
putting a carved figure-head under our bowsprits.

The battle of Salamis taught the Greeks the value
of a navy— which, by the way, they were at first afraid
to send around the southern peninsula, and so used
to haul across the isthmus of Corinth — and helped
them greatly to attain the power they afterwards had.
They improved, too, on the boats of their fathers,
extended the end platforms to meet over the heads
of the rowers so as to make a deck and give room for

more soldiers, and devised "engines" to aid
them. Their enemies were equally skilful, and the
hand-to-hand naval conflicts of those days were
horrible. From the bows protruded sharp prows
designed to strike and sink an enemy's vessel, and
over these prows projected platforms where men
stood ready to drop heavy stones or pointed masses
of iron upon the enemy's boat when it had come
within reach, to crush it; or great grappling irons
which should draw it close up so that the soldiers
could leap upon the deck of their foe and fight there.
From the masts, also, were suspended battering-rams
and other contrivances to break in the bulwarks of
an enemy; and every vessel carried catapults, or
immense cross-bows, worked by tackle and windlasses,
and which at short range shot large stones with the
force of cannon-balls. Lastly there was the dreadful Greek fire, the knowledge of how to make which
has been lost. It consisted of several liquids — the
principal one being naphtha, it is thought — which,
when mixed and exposed to the air, instantly took fire.
A vessel armed with this carried several tubes of iron
or bronze projecting from her walls, and when an

engagement came on her crew pumped through these nozzles streams of flaming liquid accompanied by dense fumes and a suffocating odor. The spouting liquid fire falling upon a man would burn him to death instantly, while the woodwork it reached was set into a blaze, so that those who escaped burning were likely to be drowned. Arrows wrapped with tow soaked in this mixture, flew through the air like meteors to set on fire any boat they lighted upon, and nothing but sand or smothering would extinguish the flames; for to throw on water was only to spread it the further. I think it a good fortune that Greek fire is lost to our knowledge, for the world could well spare so fearful an agent of harm.

After the Roman power sank there is little to tell of naval history, and nothing of commerce, until the western part of Europe began to wake up. People had sea-boats all this time, though, and were trading and fighting, and the Vikings, in their short, thick, blunt-bowed, unwieldy little vessels, had done wonders in voyaging. The natives of the British isles kept almost incessant wars with the Danes and Normans and Scandinavians back and forth across

the Channel, until the grand expedition of William
the Conqueror in 1066 settled matters somewhat.
The Normans were at that time the most powerful
people in northern Europe, and they were ambitious
to extend their rule southward.   The finest of seamen,
their fleets rounded the headlands of Spain, passed
into the Mediterranean, forced their way to Italy
and overran its ports, sailed around to its eastern
coast, and began to believe themselves masters of the
whole region.   But over there stood the refined and
beautiful city of Venice — the leader in power and
wealth of all southern Europe.   The Venetians heard
with dismay of the approach of these rough northern
strangers with their ruddy complexions, their tangled
yellow beards and their uncouth tongue.   A fleet
was quickly manned and sent to oppose them.   The
two forces met near Durazzo, in one of the hardest
sea-fights those many-battled waters ever witnessed,
and the Normans — all that were left of them —
hastened homeward and kept themselves in their own
ocean.

After that came the activity by sea and land of
the Crusades, and Venice became supreme in the

Mediterranean.   It was chiefly through the force
sent by her against the Saracens, in the twelfth century, that the Infidel fleet in the bay of Jaffa was
beaten, and that, a little later, the stronghold of Tyre
was captured.   Afterwards Genoa wrested the control
of the seas from Venice, and had her century or two of
supremacy.   But these old conflicts seem to have
produced little change or improvement in ship-building.   As before, all the long piratical wars in the
Mediterranean and along the Atlantic coast were
carried on in big row-boats, which could be managed
in a battle more easily than the clumsy, round-bodied
sail-vessels.   The fighting was all at close quarters,
and the war-vessels were often shielded on the outside with iron belts and armor, which protected them
from blows and injury from sharp rams.   To pierce
or break the vessel ; to deprive it of its rudder, its
sails, its oars ; to open in its side or at the prow a
large aperture so as to endanger its sinking, seems
to have been the tactics of every commander.   The
knights of the Middle Ages, when combatting on
shore, always fought man to man, despising anyone
who would have sought to kill the horse of the war-

rior whom he could not eject from the saddle with the
spear. At sea this was never the case. In all ages the
vessels were struck in order to overthrow the men ; and
for this reason the vessel was strengthened, cuirassed,
protected by as many towers as possible, until it
became a sort of floating fortress, bastioned, bul-
warked, defended by archers, slingers and soldiers
armed with spears, swords, axes and maces : in short,
with all the arms used in the camps or in beseiged
towns. Such were the "fleets" and tactics with
which our ancestors defended the "snug little islands,"
almost or quite up to the time of Henry VIII ; and
though merchant vessels retained the use of sails,
they were of the simplest patterns, and adapted to
little rougher work than the comparatively peaceful
waters of the Mediterranean.

In the fourteenth century we begin to hear
of a revival in the art of ship-building and the
use of sails, as, indeed, was needful if the long voyages were to be undertaken which the discovery of
the compass now rendered possible. In this revival
the Venetians and Genoese took the lead, but the
English, whose monarchs gloried in the title of

"Sovereigns of the Sea," were not far behind. There was a large variety of vessels in that day, rude though they were, and called by names we should hardly recognize now. What would you make of these lines of old Hardyorg:

> "They fought full sore upon the waters of Sayrn,
> With carrickes many, well stuffed and arrayed;
> And many other shippes great of Hispayne,
> Barges, babyngers, and galleys unaffrayed,
> Which proudly came upon our shippes unprayed."

Though the hulls of these vessels were large and tight, their shape was poorly adapted for speed or for safety in bad weather. Their decks were built up into immensely high structures at the stern and bows, after the old galley model, and to form forts for the soldiers. Our word "forecastle" reminds us of this old usage. Their masts were single sticks — not divided into top-masts — and hence were obliged to be thick and heavy; and they bore upon their summits large "top-castles" where a number of marines stood in a battle to shoot down upon the enemy's decks. This weight above, and height of surface exposed to the wind, and their rude rigging, made it impossible for them to sail safely, except with a fair and gentle wind (they never attempted it otherwise), and they were required to carry an enormous quantity of ballast. There was so little room for anything but sleeping-berths, armament and a cooking room in the war-ships, that every fleet had to take with it small vessels carrying provisions ; and the case was little better in merchant vessels.

The ships in which Vasco di Gama, Columbus, the Cabots and other explorers did their marvellous work were no better than this: strangely inefficient it seems to us, and we wonder that some of the simplest contrivances in rigging were not adopted centuries before they came into use, until we remember that it was not for long, speedy voyages that vessels were intended up to the sixteenth century or so, but simply as a means of carrying a great number of men or huge cargoes.

However, as the known world widened and trade grew, the inventions of private ship-owners continually improved the rigging, and Columbus' "carravel" had four short masts, the forward one having a square sail and the three after-masts that Mediterranean style of swinging three-cornered sail called *lateen*. At this time, too, pirates sprang up and exerted themselves to make their crafts as swift as possible, to escape their pursuers, while the regular navies, which governments now began to support, afforded opportunities to test inventions and adopt new models. To this same end also the introduction of gunpowder and cannons into warfare contributed ; for now it was no longer needful to fight hand to hand, since ships could be defeated at a distance. That gunpowder was known to the Chinese thousands of years ago, is certain ; that it was used by the Egyptians and Greeks is surmised, but it was not until about 1340 that records show equipments of cannons of brass and iron in estimates for ships, and of hand-guns for soldiers. Some of these old ship-cannon had from two to five chambers or barrels, and were covered with ornaments. Bows and arrows, cross-bows and hurling-engines remained, nevertheless, many years in company with the guns, even down to the time of Henry VIII.

From the fourteenth century, progress was rapid

towards the rigging of ships as we now see them, or
rather as they were forty years ago when sailing vessels were at the height of their prosperity, before
steamers came to do away with sails, as sails had
outstripped the old galleys.

CAT-BOAT.   FISHING SCHOONER.

1, Main-sail. SAILS: 1, Flying jib. 2, Jib. 3. Fore-sail.
      4, Stay-sail. 5, Gaff top-sail. 6, Main-sail.

The rigging of sailing vessels now is divided into
" standing " and " running " rigging; the former in-
cludes the stays to the masts, now generally made of
wire, the shrouds, and such other rope-work as is not
adjustable. A vessel looks like a skeleton when
" stripped " of all the mass of ropes, lines, halliards,

sheets and tackle which seem such a tangled maze to a landsman's eye, but are so clear and handy to the "able seaman." The sails, too, can be divided into two classes : first, those attached to a boom and gaff, or to a stay; and, second, those spread between yards which are swung across the mast, and are known as "square" sails. All the variations in shape — except the lugger sail of New Orleans — in this country can be counted in one or the other of these classes.

The styles of rigs to be seen in American waters are not many, and are easily described — at least so that you can recognize them and call them by their right names. Let us begin with the simplest :

A *sloop* has one mast and a main-sail,* and one or more jibs ; there may also be a gaff top-sail.

A *cat-boat* is a kind of single-masted boat which has no bowsprit, but has its mast stepped in the extreme prow, and only a main-sail. There is a kind of sloop-sail called a leg-o'-mutton, which is pointed at the top, has no gaff, and is seen most often in the

---

* I must refer you to the dictionary for definitions of these sails, if you don't know them now.

Connecticut sharpies, where there are often two masts rigged this way.

Of two-masted rigs, the oldest is the *brig*, which has square sails on both masts, just like the main and mizzen masts of a full-rigged ship, to be described in the next paragraph. Then there is the *brigantine*, a slight modification of the brig, and the *hermaphrodite brig*, which has schooner-rig on the main-mast and square-rig on the fore-mast. This will explain itself when you learn what a schooner is. The *schooner* is a purely American invention, and one of the greatest of all Yankee notions. It is two-masted, and the sails are alike on both, — a big squarish canvas stretched between a boom below and a gaff above, and between the gaff and the top-mast is a triangular canvas called a gaff top-sail ; the bowsprit supports one, two or three jibs. Sometimes on the foretop-mast is placed a square sail, which makes the vessel a *top-sail schooner*. Now they are building many three-masted schooners, but three-masted vessels are generally rigged as *barks* (or *barques*) or as *ships ;* or with square sails forward, and called *barkentines ;* for, though we have come to speak of any large vessel as a "ship," yet in nautical and proper language a ship is a vessel rigged in a particular way, and it is nothing else.

Masts have their proper names : the highest is in the middle of the vessel, and is called the *main-mast;* the next tallest is nearer the prow, and is the *fore-mast;* the third is in the stern, and named *mizzen-mast.* The sails and rigging take their names from the masts to which they belong, as, for example, main shrouds, mizzen shrouds, fore shrouds, mizzen-royal main-top-sail-yard, foretop-gallant-staysail, and so on. The three masts, bowsprit, yards and stays of a full-rigged ship are capable of spreading an enormous breadth of canvas — thousands of square yards ; yet in the trade-winds vessels sometimes go week after week without touching a single thread night or day. All of the sails upon the masts are square, and take their names from the sections of the mast opposite which they hang. Counting from the deck to the truck, or tiptop of the mast, these sails are as follows : On the main-mast, main-sail, top-sail (generally in two parts), top-gallant-royal and sky-sail. On the mizzen-mast are mizzen-sail, mizzen-topsail, mizzen-top-gallant,

mizzen-royal ; on the fore-mast, fore-sail, foretop-sail,
foretop-gallant, fore-royal.   The bowsprit sails are the
foretop-mast, stay-sail, jib, flying jib and outer jib.   Behind the mizzen-mast is a schooner-like sail called a
spanker ; and each of the stays running diagonally
from mast to mast bears a triangular sail known by
the name of the particular stay on which it hangs, as
main-topsail stay-sail, and so on — six in all.   In addition to all this, a little sail is sometimes set above
the sky-sail and royals, and another under the bowsprit, while out beyond the ends of the yards are
extended light additional yards carrying studding-
sails.   There may then be twenty-eight sails set at
once on a full-rigged ship, besides the studdin'-sails.
Rig the fore-mast of a three-masted vessel with square
sails, and the main and mizzen masts with schooner
sails, and you have a bark.   A *frigate* is a ship made
for war, and intended to be handled with quickness.

But the tendency is more and more toward giving
up this elaborate arrangement of lofty square sails,
and substituting three-masted schooners.   This is
due to the fact that the schooner-style will sail closer
to the wind, gives as much force in proportion as the

ship-style, while it is far less expensive to build, and more quickly and easily managed, not requiring nearly as many men, and therefore being cheaper to run as well as to set up. The schooner has worked its way, by proving its merit, well to the front in the estimation of seamen of all nations; which is why I have called it one of the greatest of Yankee notions.

VI.—WAR-SHIPS AND NAVAL BATTLES.

THE ancient mode of fighting has been told in the last chapter ; but I did not make it plain that all those old armaments were furnished as required by the various ports of a kingdom to which the ships afterwards returned, and were given back to their owners. This system was found so inconvenient that kings gradually came to build ships of their own, and to keep them at public expense ready at all times for war. In England, Henry VIII. was the first monarch to adopt this plan, and his first ship, laying the foundation of the Royal navy, was the *Great Harry*, built in 1488. For a whole century after that, England was almost continually at war on the seas with the French and Spaniards. Towards the last of this time, however, Queen Elizabeth strengthened and organized her navy very greatly, so that she was ready for the first important event in English naval history — the repulse of the Spanish armada.

This armada was a fleet of 132 large ships and many small vessels, mounting altogether over 3000 pieces of cannon. It carried 8776 seamen, over 2000 slaves to row in the galleys, and nearly 22,000 soldiers. In comparison to this, England could oppose only a miserably small fleet; and the news of the Spanish advance spread fright throughout Great Britain, and also the greatest excitement, so that, finally ninety vessels were gathered under command of Charles Howard, and cruised in the mouth of the Channel. His sub-commanders were all men whose names are famous as explorers — Sir Francis Drake, Sir John Hawkins and Sir Martin Frobisher. At the beginning of June the huge armada left Spain with pompous rejoicing; but almost at once a storm scattered it, and drove it into various ports very much crestfallen. This gave the English a chance to complete their outfit, for it was seven weeks before the armada again put to sea, and after much bad weather, appeared off Plymouth "like so many float-

ing castles." Admiral Howard let them go by, and
then sailed upon their rear. The winds were variable,
storms and calms succeeded one another, and generally only portions of the fleets were engaged at any
one time. Though so much smaller and weaker
than the Spanish ships, the English vessels sailed
better — poorly enough at best ! — were more
nimble, and could shoot squarely into the foreign
hulls, while the Spanish gun-decks were so high that
most of their cannon-balls went harmlessly over the
low-lying Englishmen. Thus for a week the two
fleets were backing and filling in almost constant
fight, which gave opportunities for wonderful feats of
single-handed courage. Then the shattered armada,
shorn of its pride, mourning a loss of forty large ships
and 10,000 men, crept back to Spain. This surprising victory was followed by many expeditions against
the Spanish coast and seizures of rich Spanish vessels.

At this time England demanded that every vessel
sailing the seas should strike its flag in the presence
of a British man-of-war, in token of submission.
This was the cause of many small fights, and began,
in 1604, the great war with the Dutch, which lasted

for fifty years, and was followed by a long series of
battles with almost all the Mediterranean ports, and
afterwards with Spain and her American islands.
This, again, was hardly over when war came on with
the French, which lasted clear through Queen Anne's
reign, and gives a flavor to so many novels written
about that interesting time.   These long constant wars,
and the incessant danger from privateers, which were
private war-vessels permitted to be equipped to range
for prizes, but not owned by the government, taught all
the European nations many lessons in sailing and
sea-fighting, so that they were a very well-built lot of
ships and fully trained crews on both sides that went
into the famous battle of Gibraltar in July, 1704.

The Dutch and English were allies then against
the French, who held the fortress not so impregnable
as now.   A heavy bombardment and assault in boats
captured the rock ; but the matter was only half done,
for there immediately appeared in the offing a large
French fleet, which was defeated only after terrible
loss of life.   Ever since then England has maintained
and continually strengthened this great rock, which
commands the entrance to the Mediterranean.

Though peace with France had been declared in
1748, war on the seas soon began again, and Spain
also became an object of attack, until in 1762
Havana was captured by a British squadron, and a
peace forced upon France and Spain by which England gained possession in America of all Canada, all
of the Southern part of the United States east of
the Mississippi, Florida, and several of the West Indian islands, with many possessions and privileges in
other quarters of the globe.   No naval victories ever
bore more fruit than did these, or were followed by
a firmer time of peace.

A little later, however, came the American war of
the Revolution; but the naval battles were few, and
well-known to most American boys.   Who has not
heard of the exploits of Paul Jones, and of the help
which Rochambeau and other French admirals gave
us?   The American navy was chiefly of that half-
piratical kind called privateers, however, and very
often came to grief.

In the midst of so many gallant exploits, and
battles of the greatest consequence, it is hard to pick
any in particular; but perhaps the most celebrated

naval encounter of the next few years was the battle
of the Nile, in which the great Nelson, who afterward
fell so gloriously at Trafalgar, commanded. A long
series of brilliant exploits had given Nelson's fame,
and the vigorous account of them he himself used to
write home helped his great popularity. In 1798 he was
a rear admiral, and was sent to the Mediterranean in
pursuit of a French fleet, which he finally cornered in
the bay of Alexandria at the mouth of the Nile. Nelson had fifteen ships, thirteen of them "seventy-fours"
(that is, carrying seventy-four guns each), and one thousand and twenty-eight guns in all; to which were op-
posed one thousand and ten guns carried by seventeen
ships, the largest of which, *L'Orient*, carried one
hundred and twenty guns. This French fleet was
anchored in a bended line and pretty close together,
only a little way from shore, the huge *L'Orient* being
in the centre. Nelson advanced with his ships, and
caused each of them to take a position as close as
possible to its particular foe, and there to anchor.
This was late in the afternoon, and the sun set upon
the sight of those thirty-two monstrous ships fastened
by their anchors in close battle array and hidden in

the smoke and flame of their incessant broadsides.
One by one the masts went down, and the hulls were
shot to pieces, until before nine o'clock at night half
a dozen ships had surrendered to the English flag,
when the whole battle was suspended at a sight which
drew all the gunners from their grimy work and
crowded the riggings of friend and foe with eager
spectators. The great flagship of the French squadron, *L'Orient*, perhaps the largest man-of-war then
afloat, was on fire. Almost in a moment the flames
mounted her tarred rigging and enveloped her massive hull. Blazing with a fierceness no land structure
would show, the remnant of her almost destroyed
crew had hardly time to throw themselves into the
water or escape in boats, when her magazines explod-
ed, and the magnificent vessel burst asunder into a
volcano of fire, to be swallowed the next instant by
the black waves.

This over, the battle was resumed, and continued
all night. In the morning a few disabled vessels of
the French were able to sail away; but though
Nelson's fleet was too badly hurt to follow them, he
had actually lost no ship except the *Bellerophon*,

which was still able to float, while several French
vessels were sunken or burned and several others
captured. The English loss of life, too, was far less
than in the French fleet, where there was not a single
vessel whose captain was not either killed or wounded. Without one exception, say historians, this
victory of the Nile is the grandest on record. The
clearness and wisdom of the admiral's plans, the fine
seamanship of the captains, the great skill of the
gunners and the splendid courage of all the men,
remain unrivalled. Now our sea battles are fought in
a different way, and these old heroic but dreadfully
bloody conflicts are things of the past.

Perhaps, though, you will think more glorious the
great battle of Trafalgar on the 21st of October, 1805.
Lord Nelson had there his flagship *Victory* carrying one hundred guns, and thirty-one other ships,
mounting altogether no less than two thousand
three hundred cannon. The enemy in this case was
the combined fleet of France and Spain, and consisted
of thirty-three vessels having two thousand six hundred and twenty guns, so that in size and number of
vessels, and in amount of ordnance, the allies had

the advantage. At daylight on the 20th the French
fleet put to sea from the port of Cadiz ; but it was
not until the next morning that they approached the
British squadrons off Cape Trafalgar. Then the
wind was so light that it was high noon before
Nelson hoisted on his ship the signals which sent
that never-to-be-forgotten message, not only to the
sailors on the other ships, but to British mariners
everywhere for generations to come —"*England expects
every man will do his duty !*"   Then the *Royal Sov-
ereign* led the way straight into and through the
enemy's line, firing her broadsides right and left
as she passed between the *Santa Aña* and the
Fougueux. Thus the terrible struggle began. Difficult to manage in the almost total calm which
always follows a naval battle, and is supposed to be
due to the shock of the air in the firing, the great
ships drifted near to one and another adversary in
turn, or became locked by some entanglement of
rigging in struggles so close at hand that the flame
of the guns burnt the woodwork of the opposite hull.
Ponderous balls went crashing through the oaken
walls, and showers of pain-dealing splinters flew to do

harm where the balls themselves had killed no one. Masts were shot away and fell headlong in a maze of rigging, or yards and heavy tackle crashed down from aloft, while all the time the thundering of the guns and the booming of broadsides rent the air and deafened the scorched and blood-spattered gunners. Nelson himself was shot by a musket-ball from the rigging of a French vessel alongside, and was carried below mortally wounded; but he covered his face and his star with a handkerchief, in order that his men should not know that their commander had fallen. Receiving reports of the battle from time to time from Captain Hardy, he continued to give orders and mourn his inactivity; but at four o'clock he became speechless, and in a few moments more was dead. Of all the noble names of naval story none are grander than Nelson's; and his last words ought to be as famous as the resolve with which he began this marvellous sea-fight: "I have done my duty — I thank God for it!"

Trafalgar was about the last of the great naval battles in the old-fashioned full-rigged wooden ships of war. Of course there was no end of hard fighting

in wooden line-of-battle ships and frigates after that
in all parts of the world ; but before many years
steam was introduced into the construction and man-
agement of ships, and iron-plating began to be used.
This produced as great a change in the methods of
naval warfare as did the superseding of the old
galleys by well-rigged sailing craft.

The guns of the old ships of war, though so
numerous, were all of small size, the heaviest of them
being only twenty-fours and thirty-twos; that is,
shooting round balls twenty-four or thirty-two pounds
in weight. The United States has lately cast some
smooth-bored cannon for land-defences, hurling a
solid round shot weighing nearly two thousand
pounds. These are the famous twenty-inch colum-
biads.

Soon after Nelson's day, the wars in the Mediterranean and in American waters, where forts had
to be bombarded and improved ships were met with,
guns of much larger calibre (that is, breadth of bore)
were made, and hurled their shot with so much
greater force that the oaken walls were no longer a
sufficient protection, and builders began to plate the

THE BATTLE OF TRAFALGAR.

hulls of their vessels with sheets of iron, whence came the name "ironclads." Next artillerymen cut spiral grooves in the bore of their cannon, like those in the barrel of a rifle, and fitted to them an elongated conical projectile — the word *ball* won't do for anything but a round shot, while *projectile* may mean anything thrown. The advantage gained by this rifling was, that a projectile (which goes straight point foremost to its mark) is called upon to punch a smaller hole than a ball of the same weight and velocity, and therefore its force can be exerted more powerfully. You can easily understand what I mean if you take a round chunk of iron and try to drive it into a board by a single blow of a hammer. You will succeed very poorly; but if you forge this piece into a nail, one strong blow will send it well into the plank. Pointed projectiles and rifled guns, then, have taken the place almost entirely in ironclad ships of war, of the old smooth-bores and round shot.

But with better knowledge of how to manufacture and how to manage large ordnance, went a constant growth in the size of guns, until now an ironclad's armament is chiefly guns having bores ten or twelve inches in diameter and carrying a projectile weight three-quarters of a ton. Such cannon would knock one of the old *Victorys* or *Bellerophons* into kindling wood in a few minutes; while there are few forts in the world — certainly not one in America — that would not crumble under their fire between breakfast and luncheon.

But as fast as the guns got bigger, the armor of vessels became thicker and the number of guns they carried was reduced. Instead of the old three and four deckers, standing as high out of the water as a church, bristling with cannons' mouths, and covered above by an enormous structure of spars and ropes and canvas, we now see long shapeless vessels with scarcely more beauty or shape about them than a canal-boat, and with only a few sails. Instead of the hundred guns there are only half a dozen, and in place of the oaken bulwarks are walls of solid iron, while the monster is driven to its position by deeply hidden engines and submerged propellers instead of working grandly into its place in the line by skilful handling of top-sails. The old romantic, picturesque glory of a sea-fight is gone ; a battle between two

modern ironclads would be much like two volcanoes
firing jets of lava at each other, for all one could see.
But we can afford to lose the romance, for the dreadful carnage of the bloody decks is also a thing of the
past; and unless some grand catastrophe occurs
which sends the whole huge ship to the bottom
before her crew can escape, the present loss of life
is small.

The most of the iron ships of war building by
England, Germany and Italy—which are the leading
nations in naval matters now — are turret ships. The
idea of a round tower of iron on deck from which
cannon should be fired, came out in our war of
the Rebellion, and produced the *Monitor*. Her
turret, however, went round on wheels, and the great
gun in it was carried along, so that instead of moving
and aiming the cannon, they wheeled and aimed the
turret. This plan has been abandoned, and the gun-
carriage is now moved within the turret, and fired
out of it as it would be out of the casement of a fort.

Of course the moving of these enormous gun-
carriages, and the handling of the heavy ammunition
in loading, is all done by steam machinery.

On the turret — sometimes a ship has two or three — and the region amidships, where (as low in down the hull as possible) the massive engines are placed, the armor is the heaviest. It consists of great plates of iron or steel moulded to the precise form of the vessels, and bolted on to the frame. These plates are as thick as possible; but when more than twelve inches in thickness is required, as is often the case, a second layer must be placed over the first.

Lastly, the prow of the ironclad below the water-line is extended out into a sharp knife or point of steel called a ram. Driven by the enormous weight of the ironclad in motion, nothing could stand a blow from this ram, and the vessel struck by it is sure to sink. Nothing, then, would be further from the thoughts of a modern captain than to allow his enemy to run into him, as used so often to be done in the old frigate fighting.

A first-class ironclad costs from $2,000,000 to $3,000,000.

VII.— ROBBERS OF THE SEAS.

A S the sea has furnished opportunities for so much
good—manly exertion, knowledge of the world,
acquaintance with people outside of one's own country, and constant wealth — so it has given a chance
for bad men to pursue their villainy, and the guarding of shore-towns and merchant vessels has always
been a part of the usefulness and duty of a nation's
naval force.

As on land there are robbers and highwaymen, so
on the ocean robber-ships have always been lying in
wait for vessels loaded with treasure, and have landed
crews of marauders to make havoc with rich seaboard
provinces.   Such robbers on the high seas were
termed pirates — so named from a Greek word mean-
ing "one who attacks at sea " — and their crime was
visited by the old laws with the most torturing meth-

ods of death. They have existed from the earliest
records of commerce, and are by no means got rid
of yet ; but they were never more daring than when
the laws against them were severest.

The first pirates who figure in history with any
great fame were those who had taken possession of
some islands in the Ægean sea, and who made forage
upon the commercial vessels plying between the
western and middle parts of the Mediterranean, and
the rich cities of the Syrian coast and the Bosphorus.
You will remember that, when Julius Cæsar was a
young man, and was making a voyage to the East
with a large number of soldiers and other persons,
he was set upon by pirates, captured and carried
away to an island where they had their homes and
warehouses.

Life was held cheap in those days by kings and
subjects both ; it need not surprise us, then, to learn
that these Greek or Byzantine pirates were accustomed to kill all their prisoners as soon as they found
nothing more was to be gained from them.   Their
usual plan was to tie them, two together, back to
back, and hurl them alive into the sea.   Some they

tortured in order to make them reveal further riches, others they retained as material for a little amusement called " walking the plank." Prisoners selected for this pleasantry were treated with the most extraordinary politeness and given the best of fare for several days, until they came to have the greatest confidence in their captors, and think them very good fellows after all. Then some fine afternoon a plank would be pushed out across the bulwark, and, amid a profusion of bows and compliments, the amazed prisoners would be invited (and compelled) " to walk home." The wretches thought it good fun to witness the terror of the men, cheated with hopes, as they fell from the end of the bobbing plank into the raging waters.

Though the pirates who took Cæsar did not know who he was, they surmised from his purple robes and large retinue that he was a person of distinction. Cæsar himself, the story says, did not seem to regard the matter as anything more than a jolly adventure, and joked the ruffians with threats of what he would do to them when he got home — he would hang 'em all, he told them. Meanwhile they treated him well, hoping to get a large ransom, and he rather enjoyed it, joining their games and amusing them by his royal airs and graces. Instead of a ransom, however, one day a large squadron of Roman galleys appeared, Cæsar was rescued, and the pirates were hanged, sure enough.

The intricate channels, many harbors and rich islands of that archipelago remained a hiding-place of sea-robbers, however, and is so yet, though every few years, from Cæsar's time till now, the kings of the surrounding countries have sent expeditions to break them up. In the sixteenth century piracy in that region was especially strong. The crews were chiefly Turkish, but the great leaders were two Roumanians, the brothers Hayraddin and Aruch Barbarossa (" Redbeard ").

Now, to understand what is coming, you must take note that in the centre of the northern coast of Africa lay the district of Algeria or Algiers, at first a sultanate of the Mohammedan Empire. When Spain expelled the Moors, and pursued her victories across the straits, however, Algeria fell under her rule, and remained so until the death of King Ferdinand in

Ye Ideal Pirate

A ROBBER SHIP.

1516. Then the Algerians sent an embassy to Aruch Barbarossa, requesting him to aid them in driving out the Spaniards, and promising a share in the spoils. He eagerly accepted this proposition, seeing more in it than his hosts had any idea of; for the moment the Spaniards had been beaten and expelled, Redbeard murdered the prince he had come there to help, seized upon the city and port for himself, and made it headquarters for that system of desperate piracy which became the dread of all Europe. These robbers of the sea called themselves *corsairs*, from an Italian word signifying a race; and they generally won, because they had the best and swiftest vessels of that time. Their flag was jet black, and their reputations became equally dark, so that even yet to call a man as bad as a Barbary pirate is to mean that he could not be much worse if he tried.

I have striven to get at the history or origin of the black flag, but have been able to discover nothing about it, except that from time out of memory a black flag has been a sign of piracy. Flags or banners, or something to serve their purpose, have always been carried by armies. In the early days of Egypt and

Israel, Babylon and Tyre, and even down to the wars
of Rome, they were made of metal, and were called
standards.   Each band, or tribe, or company of soldiers from a single region followed its own device —
usually some image or animal — cut out of thin metal
and mounted on a staff.   As tribes and armies grew
larger certain standards became prominent and lesser
ones disappeared.   Thus the eagle was the device
of the Roman Emperors; and the various eagles
which figure on the coats of arms and flags of European states perpetuate the memory of the "victorious
standards " of the Cæsars.

By and by metal pennons and banners all on one
side of the staff were used in place of the standards.
After this came cloth flags, and often these were
made of silk and were very gorgeous, particularly
those used on the old ships.   The great flag under
which William the Conqueror sailed to the conquest
of England was exceedingly costly, having been
embroidered in intricate designs by titled ladies of
his realm and presented to him.   After nations and
commerce began to take firm shape, governments
adopted national flags, and forbade private banners

except on merchant vessels within certain limits. Definite rules were made about the hoisting of flags on ships, too, and they were arranged for signalling.

Nowadays a white flag, all over the world, means truce or peace or surrender; a red flag gives an alarm, and used to be hoisted in shore-towns and on beacons to inform the people inland that an enemy was about to land and that they must hasten to the coast to repel them. A black flag means "no quarter;" that is, that all prisoners taken on both sides will be killed. As this was the general rule among pirates, and in fighting them, it came to stand for piracy. Its only ornament was sometimes a skull and crossbones.

Now let us go back to Barbarossa and his career.

Redbeard's first care was to fortify the city of Algiers; and he expended a great deal of money and labor on the perfection of the harbor, compelling all his prisoners and thousands of the citizens to work as slaves on these defences.

Meanwhile his vessels were ranging the length and breadth of the Mediterranean and cruising out upon the Atlantic, intercepting merchant ships and

fighting with vessels of war. The Spanish colonies
in America, a few years later, began sending home
immense treasures dug in the silver and gold mines
of Peru and Mexico, and extorted from the natives,
or stolen from the temples of those unhappy countries. These fleets of treasure-ships, though convoyed by war-ships, were often attacked and captured by the corsairs; and whenever it happened
that the pirates were defeated, they would land upon
the nearest unprotected coast of Spain, France or
Italy, and burn and pillage some town in revenge.
How galling this was to all merchants and travellers
we can hardly understand in these days; but so
strong were the corsairs that the fleets and armies
of various governments, and even of the Pope, which
were sent against them could not gain their stronghold or suppress their cruises, at least for more than
a short time. Not Algiers alone, but Tunis, Tripoli
and Morocco also harbored piratical vessels in
every port, and the rulers shared their spoils. This
lasted even down to the present century, and until
England got possession of Gibraltar, whereupon she
sent a large fleet to Algiers, shelled the city into

pieces, burned all the pirate ships, forced them to yield up all Christian slaves, and keep their cruisers at home. That was the end of the corsairs; but there is many a small, sneaking imitation of a pirate yet lurking round the grape and olive growing islands of that sunny sea.

Turkish and Barbary pirates were not the only ones. Though they did not go under that name, the old Norwegian vikings and the rough Norman barons were scarcely anything else in fact, as their neighboring coasts could testify; but this was away back before modern affairs began. Then, when America was discovered and the Spaniards and French began to colonize the West Indies, and to dig mines in the continents of South and Central America, a new set of pirates sprang up, than which the world has never seen worse. These were the Freebooters and the Buccaneers.

As the archipelago east of Greece had sheltered the hordes of the Turkish corsairs, so the many islands, crooked channels, reefs hidden from all but the local pilots, small harbors and abundant food of the Antilles, made the West Indies the safest place

in the world for pirates to pursue their work. To
these new and wild regions, in the sixteenth century,
had flocked bad men and adventurers from all over
the world. When the wars and their chances of plun-
der died out after the campaigns led by Cortez,
Pizarro, Balbao and the rest of the Spanish *conquista-
dores*, many ruffians seized upon vessels by force,
or stole them, and turned into robbers of the sea.
As a rule, they had farms and families on some
island, and only went freebooting a portion of the
year, at first. The large island of Hayti, or St.
Domingo, was then settled by colonists who were of
three distinct classes — farmers, hunters and cattle-
men. The last class of men spent their time in the
wild interior of the island, capturing, herding or
killing wild cattle. They came to the settlements
only now and then to get supplies, and then returned to the wilderness for several months of
absence again. Finally, a war having arisen between this and other islands, the trade of the cattle-
men was destroyed, and large numbers of them
joined the Freebooters, who then became extremely
numerous and formidable; and so largely was this

due to their new friends that they lost their old name, and were known by the name of the cattle-hunters — Buccaneer.

St. Domingo became the headquarters of the Buccaneers, but several small islands were also owned and controlled by them. They were made up of men of all nations, but were chiefly Spaniards, Dutch and negroes. They were thousands in number, possessed large fleets of ships and boats, well-armed, and had their regular chief and under-officers. The most noted, perhaps, of these chiefs was Morgan, who was an Englishman.

They had two methods of work. One was to patrol the sea in the track of vessels bound to and from Europe and Brazil or Spanish America, and seize them. Very often the crews were willing, or were compelled, to join the pirates ; but sometimes all were killed or carried into slavery. Merchant-ships, therefore, all went heavily armed in those waters, and many were the bloody battles fought.

This work, however, employed only a portion of the Buccaneers, and was too uncertain a means of wealth to suit them. They would, therefore, equip a

great fleet, enlist men under certain strict rules as
to sharing the spoils, and sail away to pillage some
coast. There was hardly an island in the West
Indies from which, in this way, they did not extort
immense sums of money under threat of destruction
of the people. The mainland also suffered from
the marauders. Great cities, like Cartagena in
Venezuela, Panama on the Isthmus, Merida in Yucatan, and Havana, Cuba, were attacked by armies
of Buccaneers numbering tens of thousands of men.
Sometimes their fortifications held good and the
enemy was beaten back; but sooner or later all
these cities, and others, smaller, were captured,
burned or partially burned, and robbed of everything
valuable that they contained.

"Why did the citizens not hide their wealth?"
They did; but the Buccaneers put to the most
dreadful tortures men, women, children, slaves — everybody—until they would tell where their money and
jewels were buried. It is sickening to read of the
crimes and suffering committed by these wickedest
of men. For years and years they were the terror of
the whole Caribbean region. Nor did their enormous

riches do them a particle of real good, for they wasted it all, the moment they got home, in wild rioting, so that the spoils earned by months of hardship, and exposure, and wounds, and danger of death, would be spent in a week of carousing. Before the end of the century, however, the combined naval forces of all the nations interested in the commerce of the new world broke the power of the Buccaneers, and their depredations ceased. Their story is one of the wildest, most romantic, but most terrible pictures in the history of the world.

For the same reason as in the case of the corsairs and the Buccaneers, the East Indies have always been infested with pirates, whose light, swift vessels could run in and out of intricate channels among the dangerous coral reefs, where government cruisers dare not follow, while the people on shore sympathized more with the pirates than with the police.

The East Indian sea-robbers, however, are, as a rule, natives of that region — Malays, Bornevans, Dyaks and Chinese, with many half-savages of the South Sea islands. This is more like a continuance of savage resistance to civilization than real piracy, since the pirates of the Atlantic are civilized sailors in mutiny against their own people and national commerce. The result is just as bad, though, for these East Indians are as bloodthirsty and cruel as can be, and if they do not kill their victims, or save them for some cannibal feast (as would probably happen in the New Hebrides and some other islands), they condemn them to a life of frightful misery. In these days of improved vessels and sea-craft, however, piracy, even in Malaya, is weak. Our consuls and government agents watch suspicious vessels; our telegraph warns the naval authorities in a moment; our steam-cruisers outspeed the swiftest craft of the black flag; our rifled guns silence their cheap artillery, and our coast surveys furnish maps so accurate that the pirate no longer holds the secret of channels and harbors where he can safely retreat. If old Redbeard should come back to life and try to be king of the seas as he rejoiced to be a couple of centuries ago, his pride would be humbled in less than a fortnight, and he would gladly return to his grave and his ancient glory.

VIII.—THE MERCHANTS OF THE SEA.

A MERCHANT is a man who buys and sells goods. When these goods are produced or manufactured within the country where they are sold, we call the buying and selling "domestic trade." When they grow or are made by one country and sold to another, we call it "commerce." Commerce, then, is trade or buying and selling between nations. Some countries produce certain things in so great abundance that they are of little or no account there, though very valuable in some other part of the world which lacks them. This "other part," however, will be sure to possess in plenty something the first region wants. An exchange of these commodities constitutes commerce, which may be done overland, as has been the method by camel-trains in Asia for unnumbered centuries, but in our age is chiefly carried on by water.

The original practise in these international trades was simple barter, that is, an exchange of a quantity of my cargo for a quantity of yours—a chest of tea for an equal-sized bale of furs, as used to be done between Russians and Chinese at the Thibetan border. Now we have money in which we estimate the value of this thing and that, independent of its bulk, and so sell what we have for cash, and buy what we want with the money. The merchants of the sea, however, have kept nearer the old method than those on land; and on wild coasts even now often make up their cargoes by exchange with the natives instead of by purchasing with money, as they would in civilized ports. This occurs most frequently among the South Sea Islands and in South Africa.

The history of shipping which you have already read in a previous chapter, will also answer as a history of early commerce. It began with the Egyptians, Phœnicians and Carthaginians, and was confined to the Mediterranean until quite modern times. Later, in the days of the Roman Empire, the trading-ships were as important to them as their soldiers; for nearly every free man was in the army, and the slaves made poor farmers. A large part of the grain, then, to supply the wants of the people had to be brought by water from Egypt, which was pretty sure to have "corn," as the Bible calls it, when the rest of the world was suffering from short crops. Large fleets of grain-ships, convoyed by armed vessels, were continually passing between the Nile and the Tiber, and so many were the risks they ran of wreck or capture, that the arrival of a flotilla with its precious freight of food was always a cause of rejoicing, at any rate among the poor people of the great city.

With that grand awakening of interest in education and industry and discovery in the fourteenth century, the city of Venice took the lead in power, and her merchants were the most enterprising and wealthy. It was the needs of commerce which urged the explorations that marked the fifteenth and sixteenth centuries, for by this time Venice had her banks — the first in the world — and her exchange on the

famous Rialto bridge; Genoa was in close rivalry;
Spain was mining immense quantities of gold and
silver in South America ; and England was coming
to the front as a maritime power. The trade with
Cathay — as all India, China, and the Oriental
Islands were called collectively — was chiefly by
caravans across the Persian deserts, and Spain,
England and Holland had small shares in it, since
the only water-route known was through the Mediterranean and Red Seas, where, between the extor-
tionate charges and stealings of the Arabs, who
carried the cargoes from vessel to vessel across the
Isthmus of Suez, and the captures by Algerine
pirates, there was little chance for profit left to the
shippers.

To western Europe, then, Vasco de Gama's discovery of the route around the Cape of Good Hope was
a long advantage, and England and Holland at least
were quick to seize it. The great " East India companies " of the Dutch and English were formed by a
group of powerful merchants in London and in
Amsterdam, who were given vast privileges by
the government in respect to trading in the East.

They equipped fleets of merchant and war vessels, established forts, carried on small wars along the Oriental coasts, and were really little kingdoms within kingdoms, because of their wide monopoly and enormous wealth. The history of the operations of these companies is full of curious interest. Their captains and supercargoes — as the men in charge of the cargo and business matters of a merchant-vessel are called — went into utterly unknown waters, and penetrated regions of land where Europeans had never been before. They brought home new facts, and specimens of human industry or natural products entirely novel. They thus enlarged the knowledge of Europe about the people and animals and plants and scenery of the East, and by taking possession of sea-ports and islands for the purposes of trade, added broad realms to their home kingdoms. Many volumes have been written in relation to the dis-coveries, adventures and great transactions of the agents of the East India companies, which were the forerunners of all our present trade with India, China and the South Pacific.

But those were slow and costly times — though full

of a romance impossible now — compared with the present. Then a voyage around the world occupied three years, and to go from London to Calcutta and back took from New Year's to Christmas under the most favorable circumstances. Now our steamers make it in less than as many days as an East India-man would have required weeks. Another important change, too, has gradually come about. Formerly, the vessels were owned almost entirely by the merchants themselves, or by a company of them. They paid all her expenses, and put into her a cargo of their own wares. They would send to China, for instance, cotton goods, household furniture, hatchets, tools, cutlery and other hardware, farming implements and fancy goods of all sorts. In return the vessel would bring silks, tea and porcelain, which would go into the owners' warehouses and be sold in their own shops. The shipper and importer and merchant were all one.

Now this is changed. The importers and merchants of London, Paris and New York are not often those who own vessels and bring their own goods. Instead of this they have agents who live permanently

TAKING PILOT IN ROUGH WEATHER; AND BOSTON HARBOR.

in each of the foreign ports where they buy the goods they want, and they hire vessels to bring them home. By the old way, the nation which had anything to sell carried it to the nation that would buy it, and brought back the best pay it could get ; now the merchants go to various parts of the world, buy their cargoes and order them sent home in substantially the same way as you go a-shopping in town.

This has brought out a new department of sea-labor, unknown, as a class, a century ago — the business of carrying goods which the owners of the vessel have no property in. In Boston, New York, San Francisco, and all other sea-board cities of this and other countries, the great majority of the vessels are owned, not by the merchants of the city, but by "transportation companies," who agree to carry cargoes at a certain rate.

In most cases these vessels run back and forth only between certain ports, and so constitute "lines," such as those between Baltimore and Rio de Janeiro. Nearly all the steamships are thus settled in their voyages, and depart and arrive with regularity once or twice a week, a fortnight, or a month. The merchant

or broker, then, who wishes to ship his goods to any
particular port, knows what vessel goes there regularly and when she will sail; or if there is more than
one line, he chooses between them carefully as to
safety, speed and cheapness.   There is thus sharp
competition between these ocean-carriers as to which
shall show the greatest advantages and transport the
most cheaply, and this is to the benefit of the public.
In addition to these there are many vessels that belong to no regular lines.   Many of them carry large
numbers of passengers also, in the most comfortable
way — another modern idea.   For people travel far
more now than they were wont to do in the times of
"good Queen Bess," or even of our own grandfathers.

All this rivalry and effort in commerce (with gen-
erous aid from scientific men and governments) have
taught navigators much about the ocean, its winds,
currents, depths and shallows, coasts and harbors. The
very shortest and safest courses are plotted upon
charts to every part of the world, and all the ships
passing to and fro between the greatest ports sail on
nearly the same courses, so that we have come to
know these well-followed though invisible tracks as

"ocean highways." A short sketch of some of these tracks will show you how they run.

The steamship lines between New York and Great Britain do not steer straight across the Atlantic, but on their way to this country keep well to the northward so as to get to the west of the Gulf Stream, and into the favorable current flowing south from Baffin's Bay ; then they skirt Newfoundland, Nova Scotia and Cape Cod. Going over, however, the steamers (and sailing-vessels too) keep much further south, and work along with the Gulf Stream as far as they can. From Europe to South America, or through the Straits of Magellan on their way to the South Sea islands or Australia (though this route is not often taken), or to the Pacific coast of the Americas, vessels keep close down the African coast, and then steer straight ahead from Guinea to Brazil and on down the coast. (Put a map before you and you will understand these courses better.) Sailing-vessels to Europe or the United States from Cape Horn, however, would swing far out into the South Atlantic to avoid heading against the southward coast-current and to get the benefit of the southeast trade-wind and the equatorial currents. From New York to the Cape of Good Hope or back, the track is nearly straight.

In the Pacific, the steamer route in summer from San Francisco might be five thousand miles as straight as a parallel of latitude, only that here, as also between New York and Liverpool, navigators adopt what is called "great-circle sailing." This consists in not heading straight for the port desired, but going to it in a curve, by which distance is saved, because the rotundity of the globe is avoided. This may be hard for you to understand, because I cannot stop to explain it here fully, but must refer you to the library.

Sailing-vessels, however, curve so far north in coming from Japan or China to America, and so far to the south going out from San Francisco, in order to get into prevailing winds and currents favorable to them, that in mid-ocean it is about one thousand miles north and south between ships outward bound and those coming home. Between California and Honolulu a steamer takes a bee-line, but sailing-vessels find it best to make detours. In summer this amounts to steering straight northward until under latitude forty degrees, before turning eastward, making more than a right angle to go around.

I have said that the finding of a sea-route to the East around the Cape of Good Hope was a great boon to Western Europe and advanced commerce. It remained so until within the last fifty years. Lately, the Corsairs being out of the way, and safety guaranteed in Egypt, merchants and sailors both began to wish they had a shorter route between England and India. Then, with immense labor and sacrifice, the canal was cut across the Isthmus of Suez, and commerce returned to its ancient channel through the Red Sea, saving thousands of miles of weary distance and much time in each voyage.

From the end of the Red Sea at Aden, the tracks of steamers both ways are straight cuts to Bombay and Ceylon, and thence straight up to Calcutta, across to Singapore or down to Australia. Except East African coast lines, no steamers go around the Cape of Good Hope from England, excepting one line to South Australia, which steers straight eastward all the way from Cape Town to Adelaide. But the Indian Ocean is so situated under the

equator, is so filled with "prevailing" winds and
currents and counter-currents, that sailing-vessels
must take very roundabout courses, and can by no
means steer the same track at all seasons of the year.

I fear you have voted these details dull; but if you
will follow them out on an atlas map which marks the
directions of the trade-winds and the ocean currents,
you will gather a great deal of interest from the sug-
gestions and information you will get.

You can by no means, however, rank all sea-faring
men as either in the naval or the merchant service.
There are other classes of industries carried on in
vessels, the fishing, for instance, which employs
thousand of men in the United States, and an
equally large proportion of the citizens of other maritime nations.   Then there are the pilots, the yachts-
men, the whalers and sealers, the coast-guard, the
life-saving and lighthouse service, the wreckers,
oystermen, and others who get their living along the
coast; besides the ship-builders, riggers, yard-labor-
ers and iron-workers on shore, the lightermen, 'long-
shoremen and warehouse hands, the brokers and
agents, supercargoes and clerks, whose daily bread

all comes directly from being busy with salt-water affairs. Add these all together, and you will find a surprisingly large number of men and families thus supported.

For the first-mentioned of these classes, pilots, space is left me for a few words in this chapter. A pilot is a man who has made himself thoroughly acquainted with certain waters where navigation is dangerous, and who directs vessels in safety through those bad places. A ship-captain may understand perfectly the proper course from one continent to another and how to handle his vessel in the open sea, but he is not expected to know every rock and sand-bar crouching under the treacherous waves, and all the twistings and obstructions of the narrow entrance into a foreign harbor. Indeed, the naval regulations will not permit captains to act, though they may *think* they know the channel, since if an accident happens when there was no pilot on board, the insurance money will not be paid.

Pilots, then, are important men, and they know it so well that they charge very high prices for their services (generally rated according to the draft of the vessel), and admit few young men to their ranks to be trained.

Their method of work is very exciting. A dozen or so together will form the crew of a trim, staunch schooner, provisioned for a fortnight or more, which can outsail anything but a racing yacht, and is built to ride safely through the highest seas. You will now and then see one of these beautiful little vessels sailing up the quiet harbor, threading her way through the black steamers and sputtering tug-boats and great ships, as a shy and graceful girl walks among the guests at a lawn party, and you know from its air as well as the big number on its white mainsail that it is a pilot boat.

But these fine schooners and the brave men they carry are rarely in port. Their time is spent far in the offing of the harbor, cruising back and forth in wait for incoming ships, and the New York pilots often go two and three hundred miles out to sea. There are other pilot-boats waiting also, and the lookout at the reeling mast-head must keep the very keenest watch upon the horizon. Suddenly he catches sight of a white speck

which his practised eye tells him are a ship's top-
gallants, or a blur upon the sky that advertises a
steamer's approach. The schooner's head is instantly
turned toward it, and all the canvas is crowded on
that she will bear, for away off at the right a second
pilot boat, well down, is also seen to be aiming at
the same prize, and trying hard to win. The man
whose turn it is to go on duty, hurries below and
packs the little valise which holds the few things he
wants to take home, and the crew's letters; if it is a
steamer which is lying there with slowly turning
wheels and signals flying, he shaves himself and puts
on a clean white shirt; but a common sailing-vessel
is not so honored.

The storm may be howling in the full force of
winter's fury, and the waves "running mountains
high," as we say, but the pilot must get aboard by
some means. It takes rough weather to make it
impossible for his mates to launch their yawl and
row him to where he can clamber up the stranger's
side with the aid of a friendly rope's-end. But often
this is out of the question. Then a "whip" is rigged
beyond the end of a lee-yard arm, carrying a rope

drove through a snatch-block, and having a bowline
at its end. The steamer slows her engines, or the
ship heaves to, and the pilot-schooner, under perfect
control, runs up under the lee of the big ship, as
near as she dares in the gale. Then, just at the
right instant, a man on the ship's yard hurls the
rope, it is caught by the schooner, the pilot slips
one leg through the bowline-noose, and a second
afterward the schooner has swept on and he is being
hoisted up to the yard-arm, but generally not in
time to save himself a good ducking in the combing
of some big roller.

Going on shipboard in this fashion is not favorable
to an imposing effect; nevertheless, the pilot is wel-
comed by officers and seamen and passengers, who
all admire his courage and trust his skill.

Now the pilot is master—stands ahead of the
captain even—and his orders are absolute law. He
inspects the vessel to form his opinion of how she
will behave, and then goes to the wheel or stands
where best he can give his orders to the steersman
and to the men in the fore-chains who are heaving
the lead. He must never abandon his post, he must

never lose his control of the ship, or make a mistake as to its position in respect to the lee-shore, or fail to be equal to every emergency. If it is too dark and foggy and stormy to see, he must feel; and if he cannot do this he must have the faculty of going right by intuition. To fail is to lose his reputation if not his life. This is what is expected of a pilot, and this is what they actually do in a hundred cases, the full details of any one of which would make a long and thrilling tale of adventurous fighting for life.

IX.—THE DANGERS OF THE DEEP.

SEAMEN, however skilful, and pilots, though never so knowing, cannot escape the incessant danger which attends a seafaring life. Experience in reading the signs of the ocean and of the skies, surveyors' charts of coasts and harbors, added to the appliances of powerful modern machinery, have lessened the perils, it is true, since the old times; yet even now ships sail proudly out of sunny havens, their topsails are watched by loving eyes till they disappear at sunset, and are never seen again. On a calm day in 1782 the great hundred-gun line-of-battle-ship *Royal George* sank unwarned into the harbor of Spithead, carrying down almost a thousand souls; two years ago the *Atalanta*, one of the finest of England's modern steam iron-clads, foundered at sea, and not a man survived. Each of these vessels was perhaps the best of its kind in the world. Every few years an extraordinary gale will prove the ruin of whole fleets in a single day, and each week, wrecks happen upon the shores of every continent, and sailors lose their lives in the most terrible manner. Old Neptune is still a match for us, when he asserts himself. Nevertheless we must go upon the restless waters, and must risk battle with their power; therefore every effort has been made by men on land to be of aid to their brethren at sea — whose perils grow greater as they approach the coast — by erecting lights to guide them to the entrance of every harbor, by marking the channels so that hidden rocks and shallows may be avoided, and by contrivances to save life and property when the fury of the gale renders seamanship useless, and the noble ship is cast away in the thundering surf of some wild shore.

Ever since men began to go to sea, lights have been placed on shore to guide them to a landing-place; but in early times these were nothing more than fires on headlands. Three hundred years before Christ, however, there was built, on an island at Alexandria, a pyramid over four hundred feet in height, on

top of which a great fire was kept burning, which, we
are told, was visible to the corn-ships going to Egypt
when forty-one miles away on the Mediterranean.
This pyramid was called the Pharos, and to-day the
French name for a lighthouse is *phare*, and the Spanish *faro*.   The rocky coasts on both sides of the British Channel centuries ago showed beacon-fires on
very dark nights.   These were generally tar-barrels,
which would burn brightly in a high wind when a fire
of sticks would be blown away.   And they were generally lighted for the benefit of the fishermen, by their
wives, without any authority from government.   It
was an easy matter to imitate such beacons, and bad
men would often erect false lights, steering by
which a ship would come crashing to sure destruction
at the foot of the crags which thrust their cruel edges
through the surf.   When the ship went to pieces, her
goods coming ashore would be seized and sold by
the *wreckers*, as these wicked people were called.
Many a fearful tradition has come down of the doings
of wreckers, not only in England and Spain, but in
America and in the East.   One of the tricks of the
West Indian pirates, when they saw a ship approach-

ing their island in the evening, was to hang a lantern upon a horse's neck, and let him graze, well hobbled, along the beach. This would appear like the rocking of a lantern on a vessel at rest; and, deceived by the hope of a safe anchorage, the stranger would only discover how he had been cheated when his keel struck the sand-bars and the pirates had begun their villanous attack.

Though scaffoldings and towers of wood, iron and stone were built here and there at especially dangerous points by governments long before the beginning of the last century, they were lighted by fires of wood or coal up to 1760, when Smeaton introduced wax candles at Eddystone. The Eddystone shoals were a group of reefs exceedingly dangerous, because they were almost invisible, and lay precisely in the track of ships bound up or down the English Channel. Two hundred years ago a lighthouse of wood and iron trestle-work was built there by Sir Henry Winstanley, and stood so well that he boasted, like King Canute, that the sea had not strength enough to throw it down. Soon after, he went out with a company of men to make repairs, when one of the worst gales in

history arose, and the morning afterward not a trace
of the structure remained. Another wooden frame
took its place for several years, but was burned.
Then the engineer Smeaton proposed to build a
tower of stone, which should take the shape of a
massive tree-trunk, with swelling base, like roots,
founded upon a level floor cut in the rock of the
reef. This stands to-day, rivalling its magnificent
neighbor on the Biscay shore opposite, the lighthouse of Carduan, which was built to support a bonfire of oak, but has remained to be lighted successively by oil-lamps, by gas-burners, and finally by
electricity. Thus, everywhere, and in all latitudes,
the beacons and wooden towers and huge pyramids
of long ago have given place to slender spires of solid
masonry, holding powerful signals perhaps hundreds
of feet above the waves, and visible as far as the
curve of the earth's surface will permit. Yet in place
of the sturdy bonfire of oak, or the huge iron cage
full of coals, there is only a single lamp, whose rays
are gathered by deep reflectors into a compact bundle
of unwasted rays, and doubled and redoubled by
rows of magnifying lenses until they can dart to the

furthest horizon in a strong beam of steady light. No longer does the mariner trust to his wife to kindle the tar-barrel to guide him home. He knows that nowhere is his government more watchful of its subjects than in its lighthouse service, and that he may trust to having that bright signal to welcome him in the darkness, as well as he can trust his own eyes to see it. The United States alone expends $2,000,000 annually in looking after her lighthouses, lightships and buoys.

Indeed, these beacons have become so thickly planted that it has been found necessary to distinguish between them in order to avoid mistaking one for another. Thus some of them are simply fixed white lights; some are white and revolve — the whole lantern on the summit of the tower being turned on wheels by machinery, and the flame disappears for a longer or shorter time; while others are white "flash" lights, glancing only for an instant, and then lost for a few seconds, or giving a long wink and then a short one with a space of darkness between. Some lighthouses show a steady red light, others alternate red and white. By these colors and vary-

ing periods of appearance and disappearance (published by the government in a book called the *Coast
Pilot*) navigators know which light they are looking
at when several are in sight, as is often the case.   The
beautiful machinery by which the light is produced,
the lenses are arranged and the revolution of the lantern effected, I cannot describe here.   If you can get
an opportunity to climb into the top of a lighthouse
and see it all, you will greatly enjoy it; and if you can
be there when a storm is raging, you will never forget
the scene.

On some especially dangerous — because hidden —
reefs or bars, like the shoals off Nantucket, or the
extreme point of Sandy Hook, it is out of the question or bad policy to erect a lighthouse.   Here its
place is taken by anchoring a stout vessel, built to
withstand the roughest weather, and arranged to carry
one or two very large lanterns at its mast-heads.
These are called "lightships," and they are manned
by a large crew of keepers who have a very monoto-
nous time of it, confined in their rolling and pitching
home with almost nothing to do.   It is never a difficult
thing to find crews for the lightships, though; there

is a class of men, much too large I think, who like nothing better than to earn their living by doing nothing, having no fear of what the French call *ennui.*

When a big storm tugs at the anchors and drives charging squadrons of water to attack the anchored vessel, the crew are kept as busy as anybody could wish, however, to save their lives and keep their beacons burning, and it often happens that they go adrift and are wrecked. Because this is liable to happen when the lightships are most needed, the government gets rid of them as fast as it can, and puts lighthouses in their place.

It is fifty years since the last Scottish coal-beacon was regularly fired for the mariners, yet the old ways occasionally crop out yet, and show that they served their purpose not so badly as we are inclined to think. Have you not heard of that noble little girl in North Carolina, whose father was a lighthouse keeper, but went ashore one night, leaving his assistant in charge? The assistant became very ill and insensible, and the little girl was unable to light the great lamps. But she saw that it was to be a tempestuous night, and

that not only her father could never sail back to her
without the light, but that vessels might strike the
rocks. So she wearily carried up the long staircases
load after load of the dry pitch-pine sticks that
had been stored for kindling-wood, and, lighting
these like torches, one by one, she held them up
inside the lantern, where their blazing was caught,
magnified and sent out to the relief of the bewildered
ships, and, best of all for the brave, tired girl, brought
her father safely to her help.

Even the electric beam from a first-class lantern
fails to penetrate a fog to any great distance; yet
when the coast is shrouded in thick mist is the
most dangerous of all times to an approaching ship.
The only way, in such an emergency, in which a warning can be given, is by sound. In many places bells
are rung; but often the point to be avoided is so
placed that the roar of the surf would drown a bell's
note, and then fog-horns are blown. These fog-
horns are of a size so immense, and voices so
stentorian, that it requires a steam engine to
blow them; and they utter a booming, hollow
blast, a dismal note as we hear it when we are

safe on the land, but sweet to the anxious captain whose vessel is laboring through the gloom under close-reefed topsails, and uncertain of her exact position. One kind of these horns is very complicated in its structure, and screeches in a rough, broken blare, a note far-reaching beyond any smooth, whistling sound that could be made. This shriek is so hideous, so ear-splitting, when heard near at hand, that no name bad enough to express it could be found ; so its inventors went to the other extreme, and called it a siren, after those most enchanting of sweet singers who tried to entice Ulysses out of his course. This name is opposite in a double sense, indeed, for the sirens of old lured sailors to wreck, while our siren hoarsely bids them keep off. Finally, buoys, which at first were simply tight casks, but now are usually made of boiler-iron, are anchored on small reefs, to which are hung bells, rung constantly by the tossing of their support ; and on other reefs, buoys are fixed having a hollow cap so arranged that when a big wave rushes over, it shuts in a body of air, under great and sudden pressure, which can only escape through a whistle in the top of the cap, uttering a long warning

wail to tell its position.   Buoys in harbors are also
made to carry lights, some by ordinary oil-lanterns,
others by having their hollow interiors filled with
greatly compressed gas, which burns in a strong globe
of glass, and cannot be blown or drenched out.

But the ordinary duty of the buoy is to mark the
line of inner channels, and by their color they tell on
which side of them the pilot must steer.

To keep the buoys all anchored, replace them if
lost, or put new ones where needed ; to visit the lightships, and carry provisions and letters to their crews ;
to see that all the lighthouses are in shape, and the
various parts of the machinery in good working order,
is the duty of an inspector, who has a certain district of coast under his care, and continually travels
up and down it in a steamer called a " tender."
Generally all the men who have anything to do with
the lighthouse service must live very lonely lives in
their towers or on their endlessly rocking hulks.
There is a small library which the government helps
circulate among them, and this library ought to be
increased, for there is no class of men in the world to
whom reading matter will prove more precious.   If

any of my readers want to give any books or files of
magazines where they will be appreciated, let them
send them to the office of the Lighthouse Board at
Washington, for this library.

Lighthouses and sirens and buoys and coast surveys are all intended to prevent shipwreck; but, as I
have said, the ocean is still supreme.   So we add to
our precautions arrangements to help those cast away.
Societies to save wrecked persons have existed in
China, it is said, for centuries, but in Europe are
scarcely over a hundred years old ; and the first life-
boat was not made until 1784.   Those European
humane societies, especially in Great Britain, placed
life-boats and gears in certain shore towns, and
organized crews who promise to go out to the aid
of any lost ship, and to take good care of the persons
rescued.   In America, however, our coasts are so ex-
tensive, and so much of the dangerous part of them
is far away from any villages, or even farm-houses,
that the government was obliged to do anything that
was to be done.   Thus came about the Life Saving
Service, as it is called, which now has its stations
close together along our whole sea-coast, and upon

the great lakes, covering more than ten thousand miles in all.

Each of these stations is a snug house on the beach, tenanted by a keeper and six men, all of whom are chosen for their skill in swimming, and in handling a boat in the surf — something every man who "follows the sea" cannot do.

During all the season, from October till May, two men from each station are incessantly patrolling the beach at night, each walking until he meets the patrol-man from the next station. No matter how foul the weather, these watchmen are out until daylight look-ing for disasters. The moment they discover a vessel ashore, or likely to become disabled, they summon their companions, and hasten to launch their boat. These boats are of two kinds. On the lakes and on the steep Pacific coast is used the very heavy English life-boat, fitted with masts and sails if necessary, and which a steam tug is required to tow to the scene of the wreck, if it is not close in shore. But upon our flat, sandy Atlantic beaches a lighter kind of surf-boat, made of cedar, can only be handled. This is built with air-cases at each end and under the thwarts, so that it cannot sink. The station-men drag it on its low wagon to the scene of its use, unless horses are to be had, and when it is launched, they sit at the six oars, each with his cork belt buckled around him, and his eye fixed on the steersman, who stands in the stern, ready to obey his slightest motion of command, for rowing through the angry waves that dash themselves on a storm-beaten beach is a matter requiring extraordinary skill and strength. Then, when the vessel is reached, comes another struggle to avoid being struck and crushed by the plunging ship, or the broken spars and rigging pounding about the hull. But skill and caution generally enable the crew to rescue the unfortunate castaways one by one, though frequently several trips must be made, in each one of which every surfman risks his life, and in many a sad case has lost it.

It is a common occurrence, however, that the sea will run so high that no boat could possibly be launched. Then the only possibility of rescue for the crew is by means of a line which shall bridge the space between the ship and the land before the hull falls to pieces. We read in old tales of wrecks of how some brave seaman would tie a light line around his waist, and dare the dreadful waves, and the more dreadful undertow, to save his comrades. If he got safely upon the beach, he drew a hawser on shore and made it fast. Now we do not ask this; but with a small cannon made for the purpose, a strong cord attached to the cannon-ball is fired over the ship, even though it be several hundred yards distant. Seizing this line as it falls across their vessel, the imperilled sailors haul to them a larger line, called a "whip," which they fasten in a tackle-block in such a way that a still heavier line can be stretched between the wreck and the land, and made fast. Then by means of a small side-line and pulleys a double canvas bag, shaped like a pair of knee-breeches, is sent back and forth between the ship and the shore, bringing a man each time, until all are saved. Should there be many persons on board, though, and great haste necessary, a small covered metallic boat, called the life-car, is sent out, into which several persons can get at once.

Such are the principal means of saving life prac-

THE DANGERS OF THE DEEP.   169

tised by the Life Saving Service; and you will believe
that they are good in device, and managed with
great skill and grand courage, when I tell you that in
1880, out of nearly two thousand persons whose
lives were endangered by shipwreck upon the American coasts, all but nine were saved.   I wish, as I
have so often wished before, that I had time to
tell you a few of the thrilling incidents of which the
surfmen are the heroes.

X. — LIFE UNDER THE WAVES.

THE ocean was the home of the first living thing that appeared on our planet, either of plant or animal ; sea-weeds and salt-water animals are found in much older rocks than any that contain the fossils of land life. Moreover, though called a " wide waste of waters," and seeming a complete desert as we gaze upon its restless surface on a dull morning, there are a greater number of animals and plants by count, and quite as large a variety, under the waves as above them, and the bottom of the sea — at any rate near its margin — is more populous than the most haunted bit of woods you ever saw.

This was more true, perhaps, of the ancient than it is of the present ocean. " The mollusks lived at that time," says one of my books, "in such serried and compact masses, that their remains have produced by

their accumulation deep strata and lofty eminences."
Is that hard to understand? Let me explain it a
trifle, and also show that not mollusks alone were in
" serried and compact masses " during the Devonian,
Silurian and cretaceous ages.

There exists in our ponds and ditches a race of
plants so minute that it requires a powerful microscope to examine them.   Under this instrument it is
seen that they have delicate, flinty shells or armor,
which is of a great variety of forms — coiled, globular,
boat-shaped, spindle-like, and so on — and always
beautifully sculptured.   These minute and beautiful
diatoms, as they are called, move about freely, and
were long supposed to be animals: now they are
known to be the simplest of sea-weeds, consisting of
only one cell.   Since life first began, these diatoms,
and other microscopic plants much like them, have
swarmed in the fresh waters not only, but in all the
oceans of the globe, furnishing food for mollusks and
all the lowly animals whose food is brought into their
mouths by the currents.   Innumerable, and as widespread as the salt water itself, every one of these
myriads of minute plants has left a record, for its

delicate, glass-like shell was indestructible, and when
the bit of life was lost, it sank slowly down to the
bottom.   What effect towards perceptible sediment
could come from a thing so small it would scarcely
be felt in your eye? One, or two, or a million, truly
would go for nothing ; but century after century,
through ages too long for us to comprehend, a steady
rain of these exquisitely engraved particles of flint
showered down upon the still sea-floor, almost as thickly
as you have seen motes in a sunbeam, until there was
deposited a layer, many feet in thickness, of nothing
but diatom-skeletons. Though this went on to a
greater or less extent everywhere in the sea, such
deposits are not now to be discovered everywhere,
because disturbing causes swept the shells away, or
broke up the floor after it had been laid down ; but
in various parts of the world to-day, you may find
wide beds of rock made up wholly of such skeletons,
soldered together into hard stone ; while in some
regions the mud of our sea-bottom appears to
consist of almost nothing else. The mighty chalk
cliffs of Great Britain and the French coast were
built up in an ancient sea, whence they have

since been lifted, in precisely the same manner.

From the simplicity of diatoms the vegetation of the sea can be traced upward through larger and more complicated kinds of plants until we reach the enormous algæ that break the gloom of black headlands by their brilliant tints, and furnish a lurking-place under their wide-spreading and dense foliage for hosts of marine animals — some hiding for safety, others to watch for prey.

Sea-weeds grow in all latitudes, even close to the pole, but mainly along the shore, for below the depth of about one hundred fathoms none but microscopic forms are known. These latter float about, of course, and many of them have been thought to be animals because they seem able to move at their own will. They come to the surface as well as haunt the depths ; and the Red Sea takes its name from the fact that a minute carmine-tinted alga occasionally comes to the surface in throngs so dense and wide as to tinge the water for miles in extent. The same thing occurs in the Pacific, where the sailors call it "sea-sawdust."

The proper home of the sea-weed, however, is a

rocky shore between tide-marks or just below them, and it is because the eastern coast of the United States is rather poor in rocks — at least south of Cape Cod — that we are poor in algæ, compared with other regions. The sea-weed has no roots, and only clings to the rock for support; shifting sand therefore would not hold it, and there are great sandy deserts under the ocean, bare of algæ, just as some land regions are sandy deserts naked of terrestrial plants.

It often happens, however, that masses of weed will be torn away from their moorings and set adrift. This does not necessarily kill them, and they go on flourishing while afloat. This is supposed to be the origin of those great areas of "gulf-weed" vegetation in mid-ocean called "sargasso seas." You will remember that a branch of the Gulf Stream, striking over towards the Moorish coast of Africa, is turned southward there, and sweeps down to the equator, then westward again. This surrounds a broad region in the middle Atlantic whose only currents go round and round like a slow whirlpool; and it is here that the gulf-weed concentrates in masses which are sometimes dense enough to impede a ship

GATHERING SEA-WEED

— Columbus reported among the wonders of his first
voyage the trouble he had sailing through it — and covers an area, between the Azores and the Bahamas, as
large as the Mississippi valley. This is the Sargasso
sea ordinarily referred to in books, but it is not the
only one. A thousand miles west of San Francisco
there is a similar collection of floating plants; and
others exist elsewhere in the southern oceans.

These floating meadows, as it were, are chosen as
the abode of a long list of animals that rarely quit
the safety and plenty of their precincts. Among
these are innumerable pretty jelly-fishes, sea-worms,
and mollusks without shells, which cling to the buoyant plants, and perhaps feed solely upon them. Here
is to be had in abundance the fairy-like, rare ptero-
pods, the richly purple janthinas with their curious
rafts of eggs, and no end of small crabs. Here a
small fish, something like a perch, spends his whole
time building a nest like a bird's in the tangled weed-
masses, and carefully guarding his treasures against
the large marauding fishes which haunt the vicinity
to the dread of its peaceful inhabitants; and here
those far-flying birds, the wandering albatross and

the petrels, hover about in search of something to capture and eat. The Sargasso sea is an extremely interesting part of the ocean, except to the luckless sailor becalmed and balked in its midst.

In favorable places a surprising variety of sea-weeds can be picked out, and books exist by which you may learn the method of classification and names of the different species. The chief of these books for America is Harvey's work, published by the Smithsonian Institution. Not only in the shape and colors of the *fronds* (as the leaf-like expansions or branching tufts of the stem are called), but in size, sea-weeds differ greatly among themselves, from the many diminutive sorts to the cable-like growths of California, which would measure a quarter of a mile in length if stretched out.

Algæ, as I have said, constitute, with very few exceptions, the whole vegetation of the salt water, together with a large part of the vegetation in fresh water ; and they serve the same useful purpose there that land-plants do for the dry parts of the globe, continually making and throwing off the oxygen which is necessary to keep the water as well as the air

pure. To this end they do a very important work.
This is not the whole of their service in ocean
matters, however. I think it can be said that if it
were not for sea-weeds, animals could not live in the
ocean, as truthfully as that if it were not for herbage, no
animals would be able to exist on land. Sea-weeds
are fed upon directly by all sorts of salt-water life, from
mollusks as big as your thumb to turtles the size of
a dining-table, and they make a shelter for thousands
of little fellows who never leave their shadow.

But this is a small part of the story. The diatoms,
and other minute plants like them, form the main portion, if not all, the food of a large number of sponges,
polyps, mollusks and other stationary sluggish creatures, that otherwise, so far as I see, would not be
able to live at all. These, in turn, are fed upon by
larger predaceous animals. Thus, though the fishes
and cetaceans * may never bite a sea-weed themselves,
they look for food to creatures that do. We may
say then that the algæ form the basis of all ocean life.

Men have been able to make marine plants of ser-

---

* I dislike a long word like this, but there is no easier English word to cover
the whales, porpoises, seals, and others of the group of marine mammals
called *Cetacea.*

vice to them also. This was more true in former
years than now.   During the last century, for example, the kelp trade was the one great industry of the
islands at the northern end of Scotland, employing
thousands of persons, and paying vast revenues to the
lordly owners of the shores.   Kelp was the ashes of
sea-weed which was burned in kilns of stone and
brick, clouding the air with huge volumes of strongly
odorous smoke.   The slow burning of the sea-weed
left the ashes fused into a solid mass, which was
broken up like stone before being sold.   In France
this was called *varec;* and in Spain, where the algæ
were mixed with beach-plants, cultivated for the purpose, and burned in holes in the ground, it went to
market as *barilla.*

In those days, kelp was the only source of the valuable alkali soda needed in manufacturing glass and
soap.   Then a French chemist discovered how to
make such soda out of common salt, and the kelp
ovens were abandoned, except a few in Scotland,
supplying the demand for iodine.   Iodine forms a
part of all sea-water and sea-weeds, and is used in
photography and in medicine.   It is a curious fact that

barbarous people have long chewed sea-weeds as a remedy in the very diseases for which physicians now prescribe the iodine extracted from those plants. Iodine is a violet dye, too, and the bluish and purple tints of many algæ, shells and sea animals appears to be due to the presence of this element.

Sea-weed and eel-grass are collected in great quantities by farmers who live alongshore in all corners of the world, as a fertilizer, especially for fruit trees. It forms an extremely good manure, because in it there is so much of the soda and lime which all plants consume; indeed, there is a kind of sea-weed growing at great depths, looking much like a coral, and called the nullipore, which takes up so much lime from the water that its substance becomes almost like stone, so that the plant retains its shape and full size when dried. Some of these nullipores are beautifully fan-shaped, scarlet or pink, and are often seen in museums, marked *Corallines*.

Cattle and horses that have been accustomed to rough pastures, like the Scotch and Irish moors, eat sea-weed and thrive on it, especially as winter fodder, and from several species are derived dishes for

our own tables.   The Irish moss, or carrageen —
which is not a moss at all, but a sea-weed — is the
most important of these, and grows on both sides of
the northern Atlantic.   In England the market supply comes chiefly from the western coast of Ireland,
while Massachusetts Bay gives America all that is
wanted.   The little port of Scituate, Mass., is the
chief point of supply, where, last year, over 400,000
pounds were gathered.   In early June, two or three
hundred men and women go to the rocks at low tide
and pick off the small brown plants, each man getting about a barrel in one day's work.   When the
tide rises, the people get into small boats and pull up
the moss with rakes.

The moss gathered each day is taken to the beach,
where a gravelly space has been prepared, and is
spread out to lie bleaching during all of the next day,
when it is taken up, washed in tubs and again spread
out.   This washing and drying in the sun is continued for seven days, by which time it has bleached
to a yellowish white.   Should a shower come, the
moss is heaped up and covered with canvas to protect
it from injury.

Besides being of value for food, carrageen serves to make sizing used by paper-makers, cloth-printers, hatters, and so on, to clarify beer in the brewery-vats, as a medicine, and to make bandoline for stiffening the hair. In cookery, jellies, *blanc mange* and various methods of boiling in milk and mixing in soups make it palatable.

Other species beside the Irish moss serve as food in Europe, generally in a raw state, often proving the only salty relish which the Irish peasant has to eat with his potatoes. One of these is the *dulse* of the Scotch, and the *dillisk* of Ireland, which also abounds in the Mediterranean, and is there made into a soup. The natives of the South Sea Islands eat algæ which is extraordinarily abundant and varied in Oriental latitudes ; and the poor among the Japanese and the interior of China, where the weed is sent dried, prize it especially, because it has a sea flavor and saves salt, which with them is a costly luxury. These people mix it with vegetables and other materials, and form thick, delicious soups and dressings. A peculiarly bad-smelling sauce, prepared from sea-weed, is among the edibles China sends to Europe as a condiment.

Along the shores from Japan to Malaga grows an alga which the natives of those coasts dry and keep as long as they please. When the substance is wanted they steep some of the dried pieces in hot water, where the weed dissolves, and then, having been taken off the fire, stiffens into glue, said to be the strongest cement in the world.

A kind of false isinglass, also, is a product of the Eastern sea-weeds, and not only helps the Chinese baker to make his pastry and confectionery, but it serves to varnish and glue thin paper and to stiffen light and transparent gauzes of fine silk used in making screens, fans, hangings, etc., so that painters can decorate them. With a poorer quality the bamboo stretchers of paper umbrellas, lanterns and various toys are smeared to give them hard and polished surfaces. In China and Japan the sea-weed is not raked up but caught by simple machines as the tide drifts it in.

Lastly, ornaments and small articles of use, like knife-handles, are made by several nations out of large dried sea-weeds; and albums of preserved fronds are one of the prettiest things to be found in a naturalist's cabinet.

XI.—SEA ANIMALS — THE LESSER HALF.

I HAVE very little space left in which to cover the great subject of ocean animals. The old idea of the ocean was that it was a vast desert; but now we know that it teems with animal life as densely as do the land, fresh waters and the air. In it began the life of this globe, for the records of the rocks show that the first animals lived in the sea, and that ages passed before any began to people the newly formed lands, and breathe the atmosphere instead of the oxygen of the water. Abundant as ocean life is now, the palæozoic seas held immensely greater hordes, and many forms which were giants compared with those of our day, just as the monstrous reptiles of the cretaceous era vastly exceeded in size the present lizards and turtles and frogs. Some of the old straight-shells, relatives of our pearly nautilus, were

185

twelve feet long; and I have seen fossil ammonites, another relative, which when alive must have been too heavy for a man to lift. The fishes, too, could tell great stories of the glory of their ancestors in size and strength and crowded hordes. Some of them wore solid coats of mail upon their great heads, and could do battle even with the huge swimming reptiles which were the dreaded tyrants of the deep.

Life in the ocean in those old geologic days was a long guerrilla warfare — every animal guarding against attack, and at the same time watching sharply for an opportunity to seize and pray upon some weaker companion. As for the foraminifera and other microscopic fellows, I have already explained how countless they were, and how their skeletons, singly invisible, have by accumulation built up great masses of rock, like the chalk-beds of England and France.

Though lessened in numbers and reduced in size, because the land has gradually won over to its side many sorts of animals which in former ages were exclusively confined to the water, and for other reasons, the sea still holds its share of

every "branch" and "class"—except birds, and
many almost claim some of them, like the albatross,
penguin and petrel—and a majority of the "orders"
of animal life. Glance at the catalogue : Foraminifers, sponges and polyps are chiefly confined to salt
water, starfishes, sea-urchins and the like, wholly so ;
mollusks (coming next higher) are principally oceanic,
and the great majority of the crabs inhabit salt water.
Among the *arachnida* (the spider order) one species,
the common " horse-shoe " of our shores, has nothing
to do with the land, except between tide-marks, and
remains as the solitary representative of an immense
and varied group which so crowded the palæozoic sea-
bottom that some rocks, for instance the limestones
of Iowa, are packed almost as full of their fossils as
a box is of raisins—I mean the trilobites. None of
the insects that I know of are truly marine ; yet some
of them are sea-faring, truly, for they spend their lives
on drifting sea-wrack, or just out of reach of the tides ;
but most of the true worms are dwellers in the mud
of the sea-bottom. I have never heard of any
land fishes yet, but I need not tell you that they
throng the fresh waters as well as the salt, and that

many species inhabit both at different seasons.

In respect to the reptiles, of which the ancient oceans contained so many gigantic and horrid types, I do not know any now that are truly oceanic, if you leave out the "sea-serpent," of which we hear so many wonderful and not quite satisfactory tales, except the turtles. Of these there are many species in various parts of the globe. You will hear of "sea-snakes" in the East Indies, but they are only certain kinds of serpents which swim well, and pass the most of their time in the salt water, as several species of our own country do in the rivers and ponds. It is in this manner, too, that we may count certain birds, such as the petrels, auks, penguins, albatrosses, frigate-birds and their kin, as belonging to the ocean. They spend all their whole days over the waves, seeking their food there, and some of them never go ashore, except to lay their eggs and hatch their young on remote rocks, resting and sleeping on the billows, or in the air over them, when not busy at their hunting. In the highest rank of all, however, the mammals, to which belong the quadrupeds and our most noble selves, several families are natives of the "great deep"—the whales, dolphins and porpoises, the seals and walruses, and the manatees and dugongs. But all these must come to the surface to breathe, not having gills like fishes, but true lungs.

Everybody has been surprised to learn during the last few years that animals are not confined to the neighborhood of the shores of the ocean alone, but exist far away from it. It is only recently that machinery has been invented for proper deep-sea dredging. Now naturalists can scrape up the soil of the bottom at the depth of even three thousand fathoms, or nearly four miles, by means of a dredge dragged by a wire rope and worked by steam machinery on shipboard. The "Challenger Expedition," which a few years ago was sent out by the British government on a cruise around the world for scientific exploration of the ocean, sank its dredges to that depth, and the information gained was very interesting.

It appears that as you go further and further away from shore, and into deeper and deeper water, the less animals are obtained, and there are very few species indeed which live on shore and also at a greater depth than about one hundred fathoms. I have already explained to you how the majority of species of marine animals are spread in limited areas of sea, beyond which other species take their place, and that it is not always easy to see why a certain sort of cowry, for example, should be found only along a particular strip of coast, when there is nothing that we can see to prevent its extending its range much further. It is believed that the *temperature* or the water is the chief fact which sets these invisible boundaries to the wanderings of animals living near the surface, only a few of which are very widespread in their distribution.

Now in deep-sea life the case is different. Here temperature cannot be of so much account, since only a short distance down, the water becomes almost as cold as ice, and preserves this uniform chill all around the globe. The life found at a great depth, too, is very widespread, instead of restricted in its range, often occurring in two or more ocean basins; but here the restriction is an up-and-down one, rather than sidewise, and the secret is found in the word *pressure*. There are very few animals able to live in the shallows and also under the enormous weight of sea water

LOBSTER-FISHING.

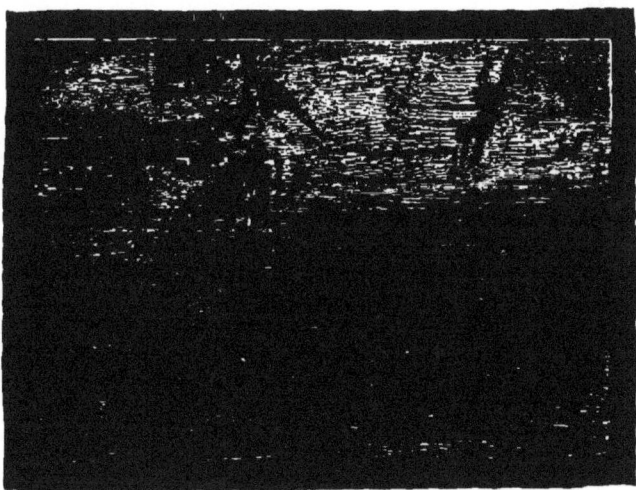

PEARL-OYSTER FISHING.

three or four miles deep. To most of them a great variation in the pressure of the water to which they are accustomed proves fatal, just as men and other animals breathing air cannot survive when they rise into the thin atmosphere beyond a certain height, or are placed where air is greatly condensed. Thus we find layers of animal life in the ocean from the shallows to the abysses.

The most striking example of this, I think, is the case of coral-reefs. The foundations are laid by the millions of minute individuals of one solid, heavy sort of coral which can grow only in pretty deep water. When these have reached their highest growth, a second kind colonizes itself upon the summit of this foundation, and carries the work a little way further. Then a third kind takes it up and brings the structure to the top, where many surface corals, corallines, and various other animals help to erect a dry reef, upon which vegetation can find root-hold, and, after a while, men may live.

There is so much to say about sea-animals, and so little time to say it, that I must only glance at the subject in a single way — utility to man.

Men make use of something in nearly every branch
of ocean-life, from humblest to highest.   The lowest
of all, as I have already said, are the foraminifers ; it
is their skeletons which make up our common chalk.
A close ally of theirs is the sponge, of which there
are a dozen or so varieties sold in the shops.   Sponges
come chiefly from the Mediterranean, the Persian
and Ceylonese waters of the Indian Ocean, and from
the gulf coast of Florida.   In the Old World they
are obtained chiefly by diving.   Men who are trained
from boyhood to this work, go out to the sponge-
ground in boats, on fine days.   Fastening a netting-
bag about their waists, and taking a heavy stone in
their hands, they dive head-foremost to the bottom —
often twelve or fifteen fathoms below — tear the
sponges from the rocks and rise with a bag-full, to be
dragged almost utterly exhausted into their boat,
often fainting immediately after.   This requires them
to hold their breath under the water for two minutes
or more ; but none but the most expert can do that,
and a diver does not live long.   In Florida, however,
the sponge-gatherers do not dive, but go in ships to
where the sponges grow, and then cruise about in

small boats, each of which contains two men. One
of these steers, while the other man leans over the side
searching the bottom. In order to see it plainly, he
has what he calls a " water-glass." This is a common
wooden pail the bottom of which is glass. Pressing
this down into the water a few inches he thrusts in
his face, and can then perceive everything on the
bottom with great distinctness. When he sees a
sponge he thrusts down a long stout pole, on the end
of which is a double hook, like a small pitchfork set
at right angles to the handle, and drags up the captive.

Having got the sponges, they must go through long
operations of rotting, beating, rinsing, drying and
bleaching before their skeletons — which is all we
want — are fit for use. Only a few out of the large
number of species of sponges are serviceable, however.

The living skeletons of the coral polyps form what
we term " corals." The round white ones and the
variously branching ones may come from any one of
several parts of the equatorial half of the globe, and
are of value chiefly as mantel ornaments. The red

coral of which necklaces and other bits of jewelry are made is procured by divers in the Mediterranean, and its gathering keeps hundreds of men busy.

Rising to starfishes and sea-urchins, I can only say that the starfishes interest oystermen because they prey upon their oysters; but in old days it was thought that medicines made out of the " stars " and " sea-eggs " were very potent in certain diseases. The trepang, which is dried and eaten by the Chinese and East Indians, belongs here too.

Coming to crabs — do we not eat them from the delicate " shedder " to the huge lobster? On the coast of Maine there are whole villages which live almost entirely by catching lobsters and canning them to send abroad. In Virginia and in North Carolina at certain seasons, hundreds of men are busy catching and shipping crabs for market, and in Louisiana large factories are devoted to canning shrimps.

This brings us to the class of mollusks, in our glance at the useful animals of the ocean; and, to prove their importance, it is enough to remind you that here belong the " shellfish " — the oyster, clam,

mussel, scallop, cockle and all the rest (not a few)
which are edible.   I find by my study of the subject,
that of oysters alone there are taken from the waters
of the coast of the United States each year almost
twenty-three millions of bushels, for which the oystermen get about $13,500,000.   That business forms
the sole support of perhaps a hundred thousand persons in this country alone.   And there are oysters
and clams and other shellfish all round the globe,
forming one of the most important of all the natural
supplies of food.   The savages knew this, as the great
shell-heaps, ancient and modern, to be found piled
upon all coasts, plainly show.

   But mollusks are not useful as food alone.   Their
shells are applied to many purposes.   We burn them
into excellent lime ; we cut them up by the million
into buttons and studs and small objects like parasol-
handles, and we polish them into ornamental shape
for our centre-tables.   In the city of Newark, I am
told, there are three factories devoted to the manufacture of shell buttons and mother-of-pearl goods
alone.   And many ship-loads of shells from the West
Indies and California come into New York annually

to supply this trade, which employs so many hands in profitable labor.

Mother-of-pearl is the bright inside surface, or *nacre*, of the large oyster which gives us pearls. This — in one or another form — exists in various parts of the world; but in America the only fishery for the pearl oyster is in the Gulf of California, and that is by no means as productive as it used to be. The season for pearl-fishing on the Pacific coast of Mexico is from June to December, but the diving can be done only in good weather, and for about three hours at the time of low water, since the tide there rises twenty feet, which would make a large dive of itself; and, besides, the currents are troublesome during high water.

At the right hour the Mexicans go out in their canoes, one man of the four or five in each canoe paddling, while the rest scrutinize the bottom. It may be rocky and weed-grown, but the water is clear, and their practised eyes detect a single round oyster where you or I certainly would overlook a dozen of them. Then down a man goes and brings up his prize, with perhaps some additional ones. Sixty or eighty feet

is not too deep for these adventurous divers, who
will stay a whole minute upon the bottom, picking up
oysters and putting them into a basket sunken before-
hand, where a quantity of shells has been discovered.
No food is eaten by these men on the day they
dive, until their labor is done.

The divers could tell us some entertaining things
about their experiences in the mysterious world
under the waves, if they had the faculty of description. Drowning is only one of many dangers that
threaten them. The warm waters in which they
work are the home of the largest and most deadly
sharks, and of various other submarine creatures
one would far rather read about than meet in his
own element. Of them all sharks are the most to
be dreaded; as a rule, however, they are easily frightened away, or can be avoided by the clever swimmer,
who quickly stirs up the mud of the bottom, and
rises in the fog before the dull shark discovers
that he is gone. The natives of the East Indies
are said to fight the sharks quite fearlessly, stabbing
them with knives as they roll over. I have read a story
to the effect that formerly the Mexican Indian divers

on our Western coast used to take down with them a stick of hard wood about two feet long and sharpened at both ends. When a shark was encountered from which they could not readily escape, they would snatch this weapon from their belts, grasp it in the middle, and thrust it dextrously crosswise into the widely distended mouth of the monster opened to seize them. To shut down his jaws upon such a skewer would undoubtedly discomfit a shark or anything else ; but when one thinks of the time, nerve, and sure aim it would require to accomplish this feat, he begins to doubt whether really it ever was tried. I advise you, therefore, to prove the story better than I have been able to do, before you pin *all* your faith to it.

All the oysters when brought ashore are opened in vats of water, and carefully examined for the pearls they may contain half embedded in their mantles; but very few reward the diver with pearls worth selling separately or except by weight. Many divers, therefore, do not themselves take the trouble of opening what they catch, but sell them unopened at a few cents a dozen, preferring the small pay and no risk to a chance for more money with so much additional work.

The round, flat, beautiful shells are all saved, and their sale (for mother-of-pearl work) brings nearly as much money into the pearl-fishing communities in the course of a season as the pearls themselves are worth.

We are not only gathering and utilizing the lower invertebrate creatures of the ocean in all these ways, but we are beginning to cultivate an artificial supply of the most important of them, such as the food-mollusks. The Romans, away back in the days of Horace, raised oysters in ponds along the Italian coast, and Eastern nations preserved the custom during the mediæval centuries when Europe was doing little except quarrelling, and making pretty pictures on parchment. Recently the French of the Channel coast took it up, and the English followed, finding that their natural oyster and mussel beds were becoming exhausted. The same fate has overtaken our own oyster-beds everywhere north of the Chesapeake, until now nearly all the oysters brought to market have been raised upon private planted beds.

An oyster-farm is conducted in two ways. One is to place upon a certain space of bottom, leased in some shallow bay, as many young oysters as it will conveniently hold. These young oysters (generally hardly bigger than your thumb-nail) are dredged in summer from certain reefs in deep water, where the oysters are never allowed to grow to full size ; and to a large extent they are brought by the ship-load from Maryland and Virginia, which have more "seed," as it is called, than they need for their own planting. These young oysters, watched against harm, and having plenty of space to grow in, come to a proper size for market in from one to three years, and are then gathered by their owners and sold.

Another method is to spread old shells, pebbles, etc., on the bottom, to which the floating eggs of adult oysters in the neighborhood adhere. This thick "catch" of seed is then taken up and respread in a thinner way upon new ground, and is allowed to grow to maturity. The oysters raised by either of these methods are of better appearance and taste, as a rule, than those that grow naturally.

Mussels, clams of many varieties, and even sponges

and peak-shells, are also cultivated to some extent,
each according to the plan its natural habits make
advisable.   In this way certain great areas of favorable ocean-bottom have become as valuable as the
neighboring shore-land, or even far more so, if you
compare, acre for acre, the yields of the crops below
with that above the water-line.

XII.—SEA ANIMALS.—THE GREATER
HALF.

IT is almost impossible for one whose home is inland, to realize to what an extent the sea enters into the industries of the people who live on its borders, and to what an extent a whole country is indebted to its adjacent ocean. This indebtedness arises in a great variety of ways, the most important of which are supplying food, and the education of a class of men whose peculiar services are often of the greatest value to their fellow-citizens.

The vegetables and lower animals of the sea which serve as food, we have already studied; those animals higher in the scale remain for this article, and they are many, — fishes, turtles, seals, walruses and whales. Each of these classes possesses a large variety and great numbers of individuals, while some of them yield other products of value beside their eatable flesh.

The right to catch fish has always been held one of the most valuable privileges a colony or nation could enjoy, and more than one war has been engaged in over this privilege ; indeed, the old French wars that tormented the New England colonies and the maritime provinces of Canada, previous to the Revolutionary war, were fought largely, if not wholly, upon this question.   By our Puritan forefathers, and by the early settlers of New York and New Jersey, the fishing along the coast was esteemed one of the highest inducements for coming to the new land, and it was talked about in a way that seems queer to us who have so many other interests ; but the forefathers were careful to have a strong clause in the Massachusetts charter as to their right over and ownership of every sort of fish and fishing, and more than once success in their boats was all that saved whole towns from starvation.   The same is true still in other parts of the world, and some savage tribes, particularly in northern regions, depend almost wholly on fish for their support.   Take away his salmon from the Vancouver Island Indian, and he would starve to death, I fear.

Sea-fishing is carried on extensively along the coasts of China and Japan, on both sides of Behring's straits, on the Californian coast, in the Mediterranean and elsewhere; but the chief fishing grounds of the world, I think, are in the North Atlantic, and particularly off the coast of Newfoundland. It is on the waters of this ocean that fishermen congregate under the greatest nnmber of different flags, and hailing from the greatest distances. To the skill and eagerness of the Norse fishermen — rovers indeed — do we owe the first voyages extending their course, by degrees, until the Orkneys, the Faroes, Iceland and Greenland successively were discovered.

The Banks of Newfoundland are a series of shoals — submerged islands, in fact — which lie off the northeastern coast of America from Cape Cod to the farther end of Newfoundland. The shallowness of the water over them makes them advantageous places for fishing, because many of the species caught, remain near the bottom, and in deep water are therefore beyond convenient reach. It is possible, also, to anchor on the Banks — a necessity for most fishing.

Cod, halibut, hake, haddock, mackerel and herring are the fishes chiefly taken in the North Atlantic, winter being the most favorable time, except for mackerel. Along our coast more southerly, menhaden, shad and mullet fishing employ great numbers of men, while on the southern coasts of Europe the catching and preserving of sardines, whitebait and anchvies is of great importance to the shore-people, beyond the ordinary fishing.

Of all these, cod-fishing takes the lead, at least in America, hundreds of vessels gathering year after year upon the Banks and loading deep with their captures. To go cod-fishing the staunchest vessels are required, for terrible storms are often encountered. These vessels are of the swiftest model, also, for they must make quick voyages to and fro, in order to save time and beat their rivals. Their rig is adapted to this purpose, and spreads almost as much canvas as a racing-yacht. The best of them all, perhaps, belong at Gloucester, Massachusetts, and are never used for any other purpose.

Formerly, all the fishing for cod, halibut and the

rest was by the hand-line, but now this is nearly given up, except in the case of mackerel, and the trawl has taken its place. The trawl consists of a strong rope, between three and four hundred feet long, having at each end an anchor and a flag-buoy. It is so arranged that when it is stretched out and anchored the line will be several fathoms beneath the surface. To this line, at intervals of six feet or so, are hung short lines, each carrying a stout hook. When the fishing-ground has been reached, the captain anchors his vessel, or, if the weather permits, he sails gently to and fro. Previously, six trawls have been baited with clams brought from home, and one put in each of the six small boats which the vessel carries. Two men now put off in each of these boats and anchor the trawls at convenient distances from each other, in such a way that the trawl-line, with its fringe of hooks, shall be stretched taut and at the proper depth. How long they stay down depends on the weather — five or six hours, or from evening until morning, is the usual length. Then the men go out, and taking up the anchors at one end, haul each trawl into the boat, coiling it in the bottom and taking off the hooks

each captive fish as fast as they come to it.

Simple as this sounds, it is terribly hard work. The trawls are heavy and stiff, and armed with dangerously sharp hooks. The busiest season is midwinter, and no dread of cold or danger must stop the fisherman who boldly ventures in his little dory into the teeth of a howling snow-storm, and fast-increasing gale, piling the water " mountain-high " about him and encasing his body in a sheet of icy spray ; this must he do in spite of discomfort and the imminent risk of death, if he would save from destruction his valuable trawls and the booty they may have hooked for him.   A fine day on the Banks of Newfoundland is a rare thing ; fog and snow and icy gales are the rule, and only the greatest courage, endurance and skill would enable a man to resist the ocean and wrest from it his self-support :

> " Brave are the hearts that man
> The fishing-smacks of Gloucester,
> The sea-boats of Cape Ann."

The thrilling tales these men can tell of escapes from death, and of exciting moments of contest with the tempest, would go far ahead of any imagined peril ever thought of ; but they cannot be written out— especially in so brief a space as these papers are allowed—without losing so much of their salt sea-flavor that they would sound very flat, I fear.

But the cod, halibut and mackerel fisheries are not the only ones that claim the steady efforts of the fifty thousand or more men whom the last census reports as getting their living by means of these industries. All along the coast from Maine to New Jersey are caught the menhaden, or pogies, or bunkers, or bony fish, as this useful little member of the "finny tribe" is called in various regions. The menhaden is shaped much like a shad, but of smaller size, rarely exceeding a foot in length. It travels about in companies or "schools," which may contain from ten thousand to five hundred thousand, and which move hither and thither in a solid mass in search of food. The food sought consists of the minute transparent animals that crowd near the surface of the summer sea, and are never suspected by any one, except, perhaps, the naturalist who, with his towing-net and microscope, captures them for his aquarium, and draws their elfin

portraits. The open-mouthed pursuit of this small
and lively prey leads the menhaden so close to the surface that the fisherman-lookout sees from his mast-
head the *flit* of their tails, and catches through the
blue wash of overlying water the gleam of their red
backs flashing beneath as they crowd one another in
rushing and wrangling haste.

The menhaden fishery is carried on in trim little
steamers, behind which are towed two yawls carrying
a great net, part of which is in each boat, a few feet
of the middle of the net stretching between them.
The moment the "color" of a school is reported by
the lookout, is a moment of immense excitement on
board the steamer. The crews rush to the boats and
pull them away toward their prize with the greatest
speed, while the steamer, left in charge of the cook
and engineer, slows her speed and gracefully swings
around to aid the boats at the proper time.

In each boat stands a man whose business it is to
pay out the net coiled in the prow ; and when the
boats have approached near enough to the menhaden — which must not be alarmed or they may
dart away by diving — the captain orders the net

thrown. Then the two men work like heroes casting
the heavy netting overboard, while the two yawls
pull away from each other, and in a circular direction, gradually surrounding the school of fish, their
track being marked by the line of cork-floats left behind and supporting the net cast overboard. It is
not until the ring of corks and netting has nearly
closed into a complete circle around them, that the
fish become really terrified and seek to escape. Then
it is too late for any but a few of them, since the boats
have drawn nearly together, and they are surrounded
by an invisible wall of netting hanging many fathoms
deep. Rushing from side to side, they push out this
yielding wall, but cannot go through its firm meshes.
By and by they would dive down and learn to go
underneath it, but the fisherman has guarded against
this by reaving through loops along the lower edge of
the net a rope, to the end of which is fastened a
heavy weight. The moment the circle of the net is
completed about the imprisoned fish, this weight is
pitched overboard. Sinking rapidly to the bottom it
pulls along with it the rope at the bottom of the net,
which, sliding through the loops, acts as a puckering-

string to draw the bottom of the net together in the centre and transform it into a huge bag or purse.

This done, the fishermen pull into their boats as much of the net as they can drag over the gunwale. This is terribly hard work. The water is often rough (I have seen them working when the crest of each wave was high enough to hide the steamer's masts only a couple of hundred yards away), and though all haul together, the struggles of the host of the fishes and the weight of the wet net allows the men to take only a portion back into the yawls. Then the steamer comes up, the net is fastened like a bag between the steamer's side and the two boats, and the menhaden are ladled out and tumbled, like a flashing shower of silver and mother-of-pearl, into the dark hold, perhaps a hundred or two hundred thousand at a single catch.

The menhaden are not eaten, but taken to factories on the beaches where they are boiled by steam in deep vats and then put through a mill that presses the rich oil out of them. The refuse of bones and scales is mixed with bone-dust, *et cetera*, and made into a land-fertilizer. Many millions of these fishes are thus

taken and used each year, but there seems no dimi-
nution of their numbers; while the cod-fish, which in
"the good old Colony times," were abundant in New
York bay, in Long Island Sound, off Cape Cod —
whence its name — and Cape Ann, have now been
driven to the far away "Banks," in an effort to escape
the incessant pursuit that follows them wherever they
may retreat. It is so likely that a few more years —
or at least a few more decades — will see their extinc-
tion, that the United States Fish Commission has
been breeding codfish artificially and seems to be suc-
ceeding in restocking a portion of Massachusetts bay.

These are only glimpses of the ways of sea-fishing
and fisher-folk — ways of hardship and peril, absence
from home, pleasant things and means of education
in books. But harder even than the lives of the
Bank and menhaden fishermen, are the duties of the
sealers and whalers.

Seals are of great variety, and of the greatest utility
to the natives of the Arctic and Antarctic regions
where they chiefly live. The Eskimos and the Alas-
kans depend upon the seals almost wholly for food
(outside of their fish-diet), clothing, fire, and lamp-

light, tools and weapons.   It is surprising how great a variety of purposes the body of the seal is made to serve; and, equally, the skill with which the active animal is killed and his carcass secured with savage weapons.

To the civilized world seals are valuable for their skins and oil, and walruses for skins, oil and ivory tusks.   The skins are employed as material for leather, and on account of their fur.   The fur-bearing seals, from which cloaks and muffs and collars are made, all live on certain islands in Alaska, where they appear every summer, departing in winter to the open sea,—nobody knows exactly where.   So easily killed were these great unwieldy beasts at these summer haunts, that the government has found it wise to make laws that no more than one hundred thousand shall be killed in one year.   This saves them from extermination, and keeps up the price of the fur so that the native population is able to support itself in steady prosperity.   Though the "fur seal," properly speaking, belongs wholly to Alaska, there is an Atlantic fur seal which formerly was much used by furriers, but is now of small consequence.

In the early spring, there come floating down with the drifting ice from the Arctic regions great numbers of seals, of three or four species, with their young. Then there are fitted out in Canada, Newfoundland, Scotland and Sweden, very strong iron steamships, manned by men who go to meet this ice-raft in the stormy seas northeast of Newfoundland, and rob it of its freight. Fastening their vessels to the pack, or working their way among the broken floes, as do Arctic navigators, the sealers brave the double perils of ocean and ice in search of the gentle little animals whose misfortune it is that they can be made useful to men. Having sighted a band huddled together with their baby seals, the crews land on the ice and attack them with clubs, one hard blow being enough to kill the innocents in most cases. Having slaughtered all the men have strength or time for, the bodies are hastily skinned in such a way that the heavy layer of fat, or "blubber," comes off with the hide, which is then taken on board the ship.

Sometimes the ice breaks up when the crews are busy in killing; or a sudden squall causes the ship to break loose and drift away; or the boats returning to

the ship are lost in fogs and snow-storms, so that the
dangers which face the sealer are far more than those
of the ordinary seaman.   Arrived at home from his
cold, perilous, exhausting and inhuman work, he sells
his sealskins to be made into leather, and the blubber
to be converted into oil.   Large fleets of staunch
steamers and hundreds of sturdy men engage in these
expeditions every year, which yield millions of dollars
to the pockets of sailors and owners; but it is all
confined to the northernmost towns on both sides of
the Atlantic.

Whaling used to be one of the very first industries
of the sea, but has now fallen into decay,— an event
brought about partly by the decrease of the whales
through incessant chasing, and partly by the discovery of petroleum, which took the place of whale-
oil for many purposes.

Old records say that whales of all kinds were com-
monly seen all along our eastern coast, and in all the
harbors.   In those days every shore town had its
boat and crew, who were accustomed to go off from
shore on a moment's notice and attack the "levia-
than."   As the increase of population and shipping

used up or drove away all the whales near shore,
ships were fitted out to go after them at a distance,
and the American colonies soon held their own with
the whalers of Europe. Boston, the Cape Cod towns,
Nantucket, New Bedford, New London and Sag
Harbor (Long Island), were the chief headquarters of
whaling, and are so yet; indeed Nantucket, New·
Bedford, New London and Sag Harbor were formed
almost wholly by the whaling and have made very
little progress since it declined.

Whaling was at its height (at least in America)
about thirty years ago, and a hundred years ago was
worth nearly as much as it is now. In 1853 the total
value of whales (oil and bone) taken was about fifteen millions of dollars; last year it was only a third
as much.

The whaling vessels were large staunch ships, and
carried crews of strong, skilful men. They would sail
on voyages lasting two or three years, and sometimes would circumnavigate the globe and return
without having touched at a port. As a rule, however, they would gain part of a cargo, and then go to
some port, ship it to London or New York, and refit

for a new voyage. The profits of a trip were thus very great sometimes, but other trips were attended by only expense and misfortune.

To the whalers we owe the discovery of many new lands and many facts in geography and navigation. At their mastheads flew the first American flags ever seen in the Pacific, and they annually added to our map of the polar regions by their adventurous struggles northward in search of new hunting-grounds. England owes to them first her colonies in the South Pacific.

The intrepid and skilful voyages, daring every fatigue and danger in the open sea, of our whalers and fishermen have been schools for the best seamen of the world. Every nation is glad to draw these sailors into their navies, and it is they who make the bravest yet most cautious captains of our merchant marine, showing to their comrades and to landsmen splendid examples of heroism and fortitude. *This* schooling I mean when I say that in its industries, we get not only food, but formation of character from Old Ocean.

NEW PUBLICATIONS.

THE PETTIBONE NAME. By Margaret Sidney. The V I F Series. Boston: D. Lothrop & Co. Price $1.25 If the publishers had offered a prize for the brightest, freshest and most brilliant bit of home fiction wherewith to start off this new series, they could not have more perfectly succeeded than they have in securing this, *The Pettibone Name*, a story that ought to create an immediate and wide sensation, and give the author a still higher place than she now occupies in popular esteem. The heroine of the story is not a young, romantic girl, but a noble, warm-hearted woman, who sacrifices wealth, ease and comfort for the sake of others who are dear to her. There has been no recent figure in American fiction more clearly or skillfully drawn than Judith Pettibone, and the impression made upon the reader will not be easily effaced. Most of the characters of the book are such as may be met with in any New England village. Deacon Badger, whose upright life and pleasant ways make him a universal favorite; little Doctor Pilcher, with his hot temper and quick tongue; Samantha Scarritt, the village dressmaker, whose sharp speech and love of gossip are tempered by a kind heart and quick sympathy, and the irrepressible Bobby Jane, all are from life, and all alike bear testimony to the author's keenness of observation and skill of delineation. Taken altogether, it is a delightful story of New England life and manners; sparkling in style, bright in incident, and intense in interest. It deserves to be widely read, as it will be.

LIFE AND PUBLIC CAREER OF HORACE GREELEY. By W. M. Cornell, LL. D. Boston: D. Lothrop & Co. Price $1.25. This is a new edition of a popular life of Greeley, the first edition of which was early exhausted. It has been the author's aim to give a clear and correct pen picture of the great editor, and to trace the gradual steps in his career from a poor and hard-working farmer boy to the editorial chair of the most powerful daily newspaper in America. The book has been thoroughly revised and considerable new matter added.

NEW PUBLICATIONS.

FIVE LITTLE PEPPERS AND HOW THEY GREW. By
Margaret Sidney. Ill. Boston : D. Lothrop & Co. Price
$1.50. Of all the books for juvenile readers which crowd
the counters of the dealers this season, not one possesses so
many of those peculiar qualities which go to make up a perfect story as this charming work. It tells the story of a
happy family, the members of which, from the mother to the
youngest child, are bound together in a common bond of
love. Although poor, and obliged to plan and scrimp and
pinch to live from day to day, they make the little brown
house which holds them a genuine paradise. To be sure
the younger ones grumble occasionally at having nothing
but potatoes and bread six days in the week, but that can
hardly be regarded as a defect either of character or disposi-
tion. Some of the home-scenes in which these little Peppers are the actors are capitally described, and make the
reader long to take part in them. The description of the
baking of the birthday cake by the children during the
absence of the mother ; the celebration of the first Christmas, and the experiences of the family with the measles are
portions of the book which will be thoroughly enjoyed. A
good deal of ingenuity is displayed by the author in bringing the little Peppers out of their poverty and giving them a
start in life. The whole change is made to turn on the
freak of the youngest of the cluster, the three-year old
Phronsie, who insisted on sending a gingerbread boy to a
rich old man who was spending the summer at the village
hotel. The old gentleman after laughing himself sick at the
ridiculous character of the present, called to see her, and is
so taken with the whole family that he insists upon carrying
the eldest girl home with him to be educated. How she
went, and what she did, and how the rest of the family
finally followed her, with the rather unlooked-for discovery of
relationship at the close, make up the substance of a dozen
or more interesting chapters. It ought, for the lesson it
teaches, to be put into the hands of every boy and girl in
the country. It is very fully and finely illustrated and
bound in elegant form, and it will find prominent place
among the higher class of juvenile presentation books the
coming holiday season.

www.ingramcontent.com/pod-product-compliance
Lightning Source LLC
Chambersburg PA
CBHW021703210326
41599CB00013B/1497